S260 Geology
Science: Level 2

The O

Block 1
Maps and landscape

Prepared for the Course Team by Iain Gilmour and Mike Widdowson

The S260 Core Course Team

David Rothery *(Course Team Chairman and Author)*

Glynda Easterbrook *(Course Manager and Author)*

Iain Gilmour *(Multimedia Development Coordinator and Author)*

Angela Coe *(Block 4 Chair and Author)*

Other members of the Course Team

Gerry Bearman *(Editor)*

Roger Beck *(Reader)*

Andrew Bell *(Author)*

Steve Best *(Graphic Artist)*

Evelyn Brown *(Author)*

Sarah Crompton *(Designer)*

Janet Dryden *(Secretary)*

Neil Edwards *(Multimedia)*

Nigel Harris *(Author)*

David Jackson *(BBC)*

Pam Owen *(Graphic Artist)*

David Palmer *(Author & Multimedia)*

Rita Quill *(Course Secretary)*

Jon Rosewell *(Multimedia)*

Dick Sharp *(Editor)*

Peter Skelton *(Author)*

Denise Swann *(Secretary)*

Tag Taylor *(Design Co-ordinator)*

Andy Tindle *(Multimedia)*

Fiona Thomson *(Multimedia)*

Mike Widdowson *(Author)*

Chris Wilson *(Author)*

This publication forms part of an Open University course S260 *Geology*. The complete list of texts which make up this course can be found at the back. Details of this and other Open University courses can be obtained from the Student Registration and Enquiry Service, The Open University, PO Box 197, Milton Keynes MK7 6BJ, United Kingdom: tel. +44 (0)845 300 60 90, email general-enquiries@open.ac.uk

Alternatively, you may visit the Open University website at http://www.open.ac.uk where you can learn more about the wide range of courses and packs offered at all levels by The Open University.

To purchase a selection of Open University course materials visit http://www.ouw.co.uk, or contact Open University Worldwide, Michael Young Building, Walton Hall, Milton Keynes MK7 6AA, United Kingdom for a brochure. tel. +44 (0)1908 858793; fax +44 (0)1908 858787; email ouw-customer-services@open.ac.uk

The Open University, Walton Hall, Milton Keynes, MK7 6AA

First published 1999. Second edition 2007. Third edition 2008.

Edited, designed and typeset by The Open University.

Printed in the United Kingdom by Cambridge University Press, Cambridge.

ISBN 978 0 7492 2679 4

3.1

BLOCK 1 MAPS AND LANDSCAPE

CONTENTS

1 INTRODUCTION TO BLOCK I

1.1 GEOLOGY IN THE FIELD AND THE LABORATORY

In their quest to understand the Earth, geologists have examined most areas of rock on the Earth's surface. They have drilled through thousands of metres of rock to collect data on everything from the rock's density to its chemical composition, spent years making careful measurements of a quietly rumbling volcano in the hope of predicting its next eruption, or collected fossils from around the world to enable the reconstruction of the Earth's past climate. Gathering hard evidence from the direct study of the Earth is the foundation of geology, and is what most geologists spend their time doing. The evidence concerning Earth processes and history is gathered from rocks, the main component of the Earth's surface. For much of the time, this is obtained by studying the rocks in their natural settings. For example, you might want to determine whether one rock is older or younger than its neighbours, whether it has been deformed, and whether the disturbance was due to some local event such as a landslide or the result of regional movements of the Earth's crust over thousands of kilometres. You can also examine the rock closely for evidence that will help reconstruct the environment in which the rock was formed; does it contain fossils, mud cracks or ripple marks, what are the minerals, how are the minerals arranged?

As we will show you throughout this Course, there are three steps that you can use in working out the geological history of an area.

1   Careful observation and collection of data. This includes mapping of the different rock types and recording their relationships in the field. Added to this is the description of the rocks, their properties as they are observed when visible in the field and hand specimen, and measuring their orientation with respect to each other. More detailed work is undertaken in the laboratory, where 'thin' (30 μm) sections of rock are examined under the microscope for identification of minerals and rock textures (relations between the various minerals).

2   The evidence gathered is put into a regional context and compared with additional information on the area, collected by others. You are seeking the answers to specific questions: was the rock formed under water, was that water moving, what depth was the water? For different rock types, different questions will arise from the evidence you have gathered: was the rock formed from a volcanic eruption or was it emplaced as a molten rock deep within the Earth's crust?

3   The geologist is often compared to a detective visiting the scene of a crime millions of years after it occurred. Once you have gathered your evidence and interpreted it in terms of the geological processes involved, then you may be able to reassemble a particular episode or event from the Earth's past to reconstruct the environment in which the rocks were formed.

1.2 THE DEVELOPMENT OF GEOLOGY AS A SCIENCE

Geology is a relatively young science, having begun to evolve into something like its modern form in the late 18th century through the efforts of scientists such as James Hutton of Edinburgh and William Smith of Bath. Hutton's ideas may be expressed in three closely related concepts.

1   Rocks contain evidence of processes which we can see operating today. He expressed this observation in the maxim that 'the present is the key to the past', which is encapsulated in the **Principle of Uniformitarianism**.

2   Observations reveal that many geological processes, such as **erosion** and sedimentation, cause change slowly in small increments; however, the sum of these changes is enormous. He inferred from these slow rates that very large spans of time must be involved, orders of magnitude longer than all of human history.

3   The Earth is a dynamic planet whose surface is in a continual state of change, its materials being continuously cycled and recycled.

Hutton's first concept – uniformitarianism – was directly opposed to the religious ideas of the time, which held that geological history could be explained by a series of sudden violent events such as the Biblical Flood, a position called catastrophism. Hutton's second concept was also in conflict with religion, since a leading proponent of catastrophism had calculated from 'clues' in the Bible that the Earth was created in 4004 BC.

The contribution of William Smith, a surveyor and canal engineer by trade, was to make careful notes of the fossils he collected in various layers of rock over widely separated tracts of the English countryside. On the basis of his observations, Smith proposed the **Principle of Faunal Succession**: that particular groups of fossils characterized each layer, and that the same succession of fossil groups from older to younger rocks could be found in many parts of the country (Figure 1.1).

Figure 1.1   The Principle of Faunal Succession: the similarity of some fossil assemblages in different areas can be used to determine the relative ages of strata. Names refer to periods of geological time that will be introduced shortly.

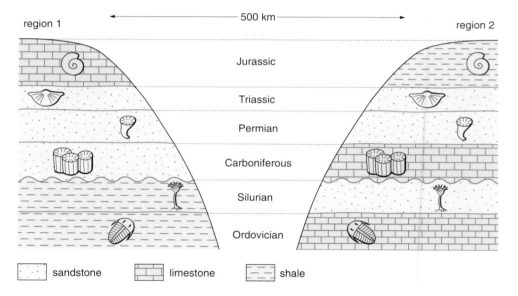

❑   Two rock successions separated by thousands of kilometres contain identical assemblages of fossils. What does the principle of faunal succession imply about their relative ages?

◼   The principle of faunal succession is that fossil organisms succeed one another in a definite and recognizable order. Rocks containing identical fossils must therefore be identical in age.

Both Smith and Hutton knew that **sedimentary rocks** were deposited on the sea-floor in layers, known as **strata**, parallel to the Earth's surface, so that the oldest layer is at the bottom with successively younger layers resting on top; this is called the **Principle of Superposition**. The three separate principles of uniformitarianism, faunal succession and superposition were brought together and further developed by Sir Charles Lyell in his book *Principles of Geology* which was published in the 1830s. This book popularized the work of Hutton and Smith and led to a widespread interest in geology. These early geologists established the essential nature of geology as an observational science, with intensive collection and description of minerals, fossils and rocks.

1.3 GEOLOGICAL TIME AND MAPS

1.3.1 ESTABLISHING GEOLOGICAL TIME

The pioneering geologists of the 19th century had noted the evidence they found of coral reefs apparently now stranded in the hills of northern England, and of complex and sometimes unusual life forms that had no living counterpart. They knew the rocks possessed a long story within them, but they did not know how long. One of the most challenging problems facing the new science of geology was to calculate the age of the Earth. Hutton was able to determine the sequence in which rocks formed – their relative age – by examining the physical relationships among rocks. However, to compare the ages of a rock succession in the British Isles to one in North America, we need to use Smith's principle of faunal succession and identify the fossils present. By comparing and correlating rock strata across the world, geologists have been able to construct a relative geological time-scale, relative because it indicates the sequence of events and not the actual time at which they occurred. If fossils succeed one another in a recognizable order, each species must have existed for a certain time interval and then become extinct. The first and last appearance of each fossil therefore represents a fixed range of geological time.

Much of the work to establish the geological time-scale was done in the 19th century as a result of painstaking records compiled by the geologists of the time and it is one of the great achievements of science. As you can see in Figure 1.2,

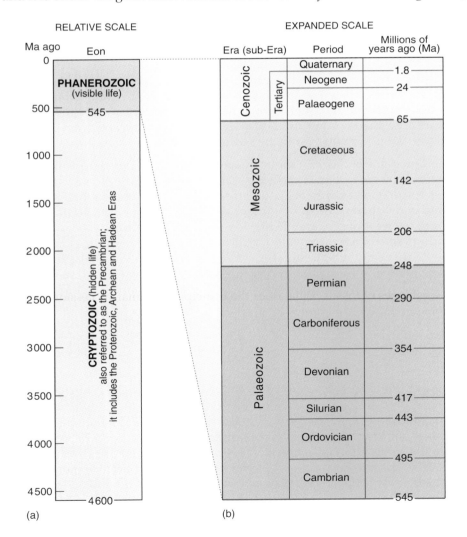

(a)   (b)

Figure 1.2 Stratigraphic column for the whole of geological time: (a) to true scale and (b) expanded scale since the onset of the Cambrian. The scale is divided into Eons, Eras and Periods. In (a), note the length of the Cryptozoic, about 90% of all geological time.

the oldest division of the time-scale is listed at the bottom, with successively younger ones on top. The scale is divided into **Eons**, **Eras** and **Periods** which are shown systematically as time-divisions and arranged in chronological order with the oldest period at the bottom and the youngest at the top to produce a **stratigraphic column**, just as sedimentary rocks of these ages would be stacked now if they had been left undisturbed since they were laid down. It is essentially Smith's stratigraphic succession for the whole of Britain, and we will examine this geological time-scale in more detail in the next Section.

Notice that Figure 1.2 also shows ages as numbers. These ages, given in millions of years, are a later addition that came in the 20th century. The various methods used in the 19th century to estimate the spans of time represented by different thicknesses of rocks produced the conclusion that the Earth was at least several hundred million years old. However, in 1865 the physicist Lord Kelvin determined the age of the Earth to be between 20 and 40 million years. His calculations were based on the time needed for an originally molten Earth to cool. The subsequent application of the rates of decay of radioactive isotopes led to radiometric dating of rocks, and these dates showed Kelvin's estimate for the age of the Earth to be much too small. The main reason is that the Earth is not simply cooling but heat is also being produced by radioactive decay, thus slowing the overall cooling. As Kelvin's work was done long before radioactivity was discovered, he could not possibly have allowed for it in his calculations. From time to time, new results cause key dates in the geological record to be revised by a small amount (which explains why you may find dates differing from those we use in this Course), but the essence of the time-scale in Figure 1.2 is not generally in doubt.

The true scope of geological time that the absolute dates revealed surprised geologists. Analysis of meteorites, believed to be the oldest surviving material in the Solar System, indicate that the Earth was formed some 4600 million years ago. Further work has revealed that much of the fossil record on which the relative time-scale is based spans only the past 540 million years, with older fossils being both rare and often poorly preserved. The expanse of time represented by an age of 4600 million years is difficult to grasp within human experience and is perhaps best treated by analogy.

If we compress the entire 4600 million years of geological time into a single calendar year, then the oldest rocks we know of on Earth, the 3800 million-year-old Isua Formation in Greenland, date from February. Life probably first appeared in the sea at some point during early March though it initially had no hard parts to be fossilized, and no life appeared on land until late November. The widespread forests and swamps that formed the coal deposits of much of Europe flourished for about four days in early December. The dinosaurs were in their heyday in mid-December but disappeared at the time a large asteroid or comet hit the Earth around lunchtime on 26 December. Human-like creatures appeared in the late evening of 31 December and the last Ice Age ended around 1 minute and 15 seconds before midnight on that day. Columbus arrived in America some three seconds before midnight and James Hutton established his Principle of Uniformitarianism slightly more than one second before the end of the year.

Recognition of the immense span of geological time is perhaps geology's most significant contribution to modern science. Over the past 400 million years, plate tectonic movement has resulted in the progressive northerly drift of the British Isles region from some 30° south of the equator to around 55° north. As a result, the rocks of Britain reflect a sometimes turbulent geological history from ancient mountain ranges and the remains of long extinct volcanoes brought about by collisions between plates to limestones laid down in tropical coral seas and sandstones, the remnants of desert sand-dunes (Plate 1.1 in the centre of this Block).

Activity 1.1*

We have prepared a video to help you grasp some of the concepts of geological time (video sequence *Geological time*) on DVD1, which you should now view as part of Activity 1.1. Please refer to Activity 1.1 in the Workbook to begin this exercise.

1.3.2   A SIMPLE GEOLOGICAL MAP OF BRITAIN

We shall begin our study of the maps themselves by looking at the small Geological Map of Britain and Ireland shown in Plate 1.2. This map has been compiled by reduction and simplification of data from many maps covering much smaller areas in great detail. You will see from the key to Plate 1.2 that the sedimentary rocks are divided into four main groups – **Cenozoic**, **Mesozoic**, **Palaeozoic** and **Upper Proterozoic** – which correspond to particular spans of time in the stratigraphic column of Figure 1.2 (numbers in bold type to the right of each column in the key to the map give ages in millions of years). The first three groups are called Eras of geological time and together constitute the **Phanerozoic** Eon. Time before the Phanerozoic is divided into three other Eras (not shown on Figure 1.2), the **Proterozoic**, the **Archean** and the **Hadean**, which together constitute the **Cryptozoic** Eon (or alternatively the **Precambrian**). The term Upper Proterozoic on the map refers to the most recent part of the Proterozoic Era. The Eras of the Phanerozoic are themselves divided into Periods whose names are given in bold type next to the colour key in Plate 1.2 (e.g. Permian, Cambrian). For the sedimentary rocks, each colour represents rocks formed during a particular Period.

The stratigraphic column is one of the most important features of the key on any geological map; it is always arranged in **stratigraphic order**, with the oldest rocks at the bottom. You will see that there are two important departures from this principle on the key to the simple map. The **metamorphic rocks** are divided into two main groups which do not appear to relate to periods in the stratigraphic column, and the **igneous rocks** are not classified by age, but are divided into **intrusive rocks** and **volcanic rocks**. Igneous rocks are divided into only two categories to simplify the map, and not because their ages are unknown. The metamorphic rocks are divided on the basis of the age of the metamorphic events which have affected them, and these events do not necessarily correspond exactly to particular periods.

You should now have a look at the geological map of the British Isles in Plate 1.2 and then answer the following questions which are intended to consolidate your grasp of the relationship between the stratigraphic column of a geological map and the map itself.

> **Question 1.1**   Using the key in conjunction with the map, where are the oldest and youngest rocks in the British Isles found?

> **Question 1.2**   Imagine that you are travelling in a perfectly straight line from London to Edinburgh. Write down in sequence the ages of the rocks (i.e. the periods) over which you would pass on this journey. (Ignore the red and purple patches of igneous rocks in the northern part of the journey.)

From your answers to Questions 1.1 and 1.2, you should now be aware that not only are the oldest rocks in the north-west and the youngest in the south-east but also, for much of the intervening distance, the rocks become progressively older from London northwards. The progression is not usually as regular as along this (carefully chosen) line from London to Edinburgh, but the general trend in Britain is always towards older rocks to the north and west.

* Activities in this Course require you to break off study of the text and do something in your *Workbook* for Block 1. Icons in the margin indicate the material you need to perform the Activity. Video sequences are located on the DVDs along with other DVD Activities. Comments on each Activity appear at the end of the *Workbook*. You should read these *after* attempting the Activity and *before* continuing with your reading of the text.

Question 1.3   Where are the areas of metamorphic rocks in the British Isles?

Question 1.4   Where are the areas with intrusive and volcanic igneous rocks? Are they associated mainly with metamorphic rocks or with sedimentary rocks?

For Questions 1.5 and 1.6, you will have to use the Plate 1.2 map in conjunction with the outline map below.

Question 1.5   On the outline map (Figure 1.3a), which has a number of unlabelled geological boundaries, use information from Plate 1.2 to shade in the main areas of rocks which together belong to the Palaeozoic Era, the Cryptozoic Eon (i.e. Archean and Proterozoic Eras) and/or consist of igneous and metamorphic rocks.

Figure 1.3   Outline map of Britain showing (a) some major geological boundaries, and (b) higher ground.

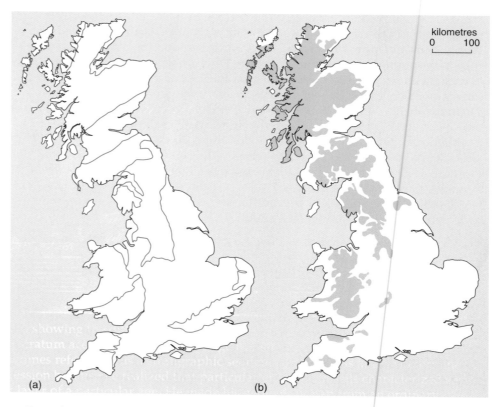

Question 1.6   Look at the areas you have indicated for Question 1.5. Is there any relationship between those areas and the high ground shown on Figure 1.3(b)? If so, why do you think there is a connection?

The last important point we wish to make about the simplified geological map of Britain is that each period, although represented on the map by a single colour, may include a variety of rock types. For example, the Silurian (lilac) includes shales, mudstones and limestones, and the Jurassic (olive-green) includes limestones and clays. This simplification is necessary to be able to show the geology of Britain on such a small map. If you have been unable to resist opening your Ten Mile Map sheets, you will have noticed that they show a great deal more detail within periods, often with one colour for each type of rock or group of closely related rocks. On larger-scale maps of smaller areas, it is possible to show even more geological detail: at the scale of your Moreton-in-Marsh map in the Home Kit, it is possible to distinguish different beds of limestone. Thus, the key on that map is related not only to particular time divisions but also to particular beds of rock.

You should now familiarize yourself with the names of the eons, eras and periods from Figure 1.2 and/or Plate 1.2, though you do not need to remember any of the numerical dates.

1.4 GEOLOGICAL MAPPING

1.4.1 HISTORY OF GEOLOGICAL MAPPING

Cartography, the science and art of drawing maps, is one of humankind's oldest activities. Some of the earliest-known maps are preserved on ancient mudstone tablets from Babylonian times. Apparently, from very early in our history, there was a need to represent the relative positions of known places. The development of geological maps could not really advance until base maps existed, on which the essential geographical and landscape features could be shown. Accurate maps did not generally become available in Britain until about the end of the 18th century.

Hills, valleys and other undulations of the land surface, are known as **relief**, and maps showing these details are called **topographic maps**. One of the problems that early map makers faced was how to portray the relief of the land surface on a flat piece of paper. On the earliest maps, hills were shown in a very diagrammatic way as little 'cones' dotted about the map to represent where the top of the high ground lay (Figure 1.4).

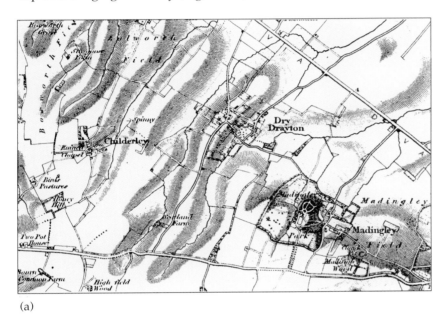

(a)

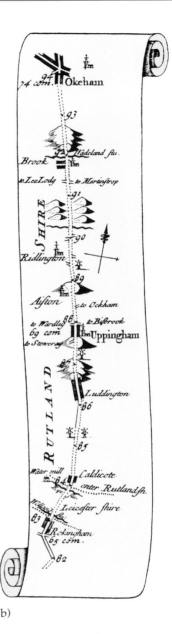

(b)

Figure 1.4 Relief on early maps. (a) Detail from the Cambridge sheet of the first Ordnance Survey One-Inch Survey of England and Wales (1836). Scale 1 inch to 1 mile. (b) Detail from a map of the road from London to Oakham (spelt 'Okeham' on the map) in Rutland drawn by John Sennex (1719). Scale approximately 0.5 inch to 1 mile.

Large-scale plans of mines, together with sketches showing geological data, were produced throughout the 18th century and Figure 1.5 (overleaf) shows an early example of a geological sketch map of buried strata in the Somerset coalfield. These maps represented the first attempts to produce **geological cross-sections** giving the geologist's interpretation of the rocks below the Earth's surface. In the 1790s, William George Maton, an undergraduate at Oxford, toured the western counties of England and made the first regional geological map of any area in Britain. On his map he indicated areas of different rock types by line shading (Figure 1.6).

However, it was William Smith who, more than anyone else, really established the idea of the stratigraphic column in England and provided a framework that allowed geology to develop very rapidly as a science in the early 19th century. His work as a surveyor began on the Somerset Coal Canal near Bath where the rocks are exposed in the rather hilly landscape, as well as in quarries worked for building stone. In 1797, he made a small geological map of the country around Bath and he soon discovered that he could trace the strata of that area farther across England. We know that Smith had a copy of the cross-sections of the Somerset coalfield (Figure 1.5); however, it seems that he had the quite original

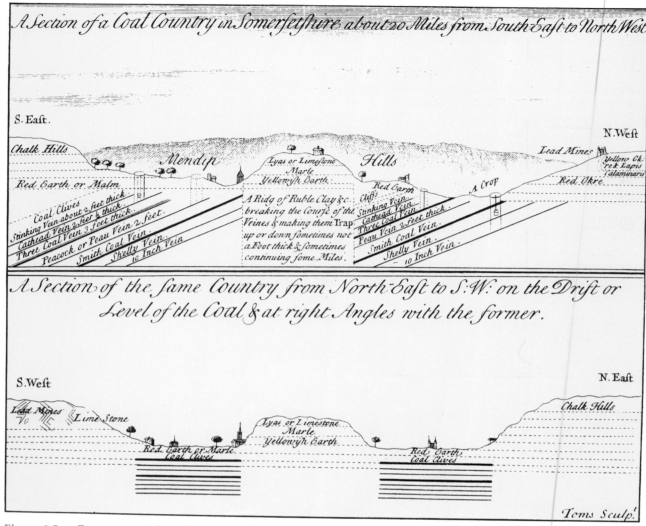

Figure 1.5 Cross-sections of the Somerset coalfield, published in 1725.

idea of showing the different strata on a map. He did this by assigning a colour to each **stratum** according to its relative age in a stratigraphic succession (also sometimes referred to as a stratigraphic sequence). Smith was able to determine the succession because he realized that particular groups of fossils characterized each rock layer of a particular age. He made his geological map from an ordinary topographic map by colouring in the area where each stratum was found on the ground.

By 1801, Smith had made an outline map which he called the 'General Strata of England and Wales'. He had traced the strata from around Bath to the north-east along the line of the Cotswolds as far as the Yorkshire coast. As he increasingly accumulated geological information, he realized that if he wanted to produce a large-scale geological map of the country, he would require a suitable base map. Fortunately, the publisher J. Cary in the Strand, London, was able to supply just that. Cary's maps allowed Smith to locate himself accurately, hence helping with marking geological boundaries. Smith's great map, an extract of which is reproduced in Plate 1.3, measured almost 3 m × 2 m in 15 sheets and was published in 1815. We know from the surviving copies that probably more than 400 were produced over about five years, and we can trace modifications to the map during that period. Smith had his own unique method of colouring the strata: instead of applying a uniform colour to each stratum, he faded the colour away from the base of each bed. This gave his maps a quality of their own and helped considerably to emphasize the three-dimensional nature of the strata. William Smith went on to produce separate maps of about 21 counties on an even larger scale, all on Cary base maps, and we can regard him as the originator of the British geological map as we know it today.

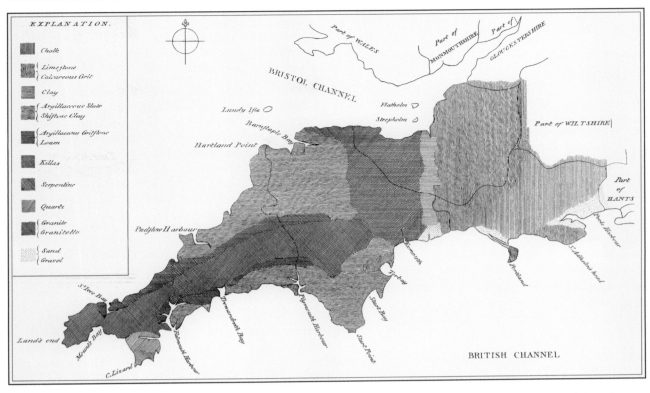

Figure 1.6   Maton's 1797 map of south-west England.

1.4.2 THE ORDNANCE AND GEOLOGICAL SURVEYS

Cary's maps were a considerable improvement on those of earlier cartographers, but all the survey work and printing relied on the initiative of individuals. In 1791, the Board of Ordnance was founded under the control of the British Army. The immediate need was for more accurate and systematic knowledge of the nature of the landscape for the military during the prolonged war between Britain and France.

The work of the Board of Ordnance introduced a more systematic approach to mapping: a detailed survey of the country was carried out and heights of prominent hills were related to sea-level. The most important step of all in topographic mapping soon followed: the representation of relief on maps by **contours** drawn at regular height intervals replacing the method of showing relief by hachuring which had persisted well into the 19th century. Each contour on the map represents the intersection between the topography and an imaginary horizontal plane at a specific height above mean sea-level. Mean sea-level is used as the reference plane, and is known as **ordnance datum** (OD). Later advances in cartography included the standardization of map scales, and the establishment of the **National Grid** system of reference for Britain that enables particular features to be located easily (see Box 1.1 overleaf).

With the development of good topographic maps, the way was open for the preparation of accurate geological maps like the ones which accompany this Course. The process began in Ireland in the 1820s, when surveyors of the Board of Ordnance started to add geological information to their maps. However, before long it was realized that surveyors could not adequately undertake the geological work since they were primarily concerned with preparing topographic maps. In 1835, Henry de la Beche was the first person appointed to plot the geology on the newly completed Ordnance maps of Devon and Cornwall. Subsequently, the Geological Survey of Great Britain was established as a parallel organization to the Board of Ordnance, which itself was renamed the **Ordnance Survey** (OS). The Geological Survey became part of the **Institute**

Box 1.1 The National Grid

Any point in the UK can be located on a map by means of its **grid reference**. If you have little or no previous experience dealing with Ordnance Survey (OS) maps, then you should read this Box as you will need to understand how to find a locality from its grid reference in the rest of this Block. The National Grid is in effect a grid overlain on a map of Britain (Figure 1.7). This grid is divided into 100 km grid squares, each grid square identified by two letters and two numbers (in brackets). For example in Figure 1.7a, the grid square that encompasses the area around Land's End in Cornwall is SW (10) which can be abbreviated simply to SW, while the grid square that encompasses Edinburgh is NT. You may have noticed that the country appears to be split in two with most of the 100 km grid squares in Scotland and northern England starting with the letter N (except for Orkney and Shetland which begin with H), while those in southern England start with S or T.

❑ What is the distance in kilometres between the bottom of the southernmost S grid squares in Figure 1.7a and the bottom of the southernmost N grid squares?

■ This distance is 500 km.

The east–west running lines of the grid are referred to as **northings** while the north–south running lines are called **eastings**. The grid line that separates the N grid squares in the north from the S grid squares in the south is therefore referred to as the 500 km northing since it is 500 km north of the bottom of the grid.

The 100 km grid squares are further divided into smaller squares by grid lines at a 10 km spacing, each numbered 0 to 9 from the south-west corner in an easterly (left to right) and northerly (upwards) direction. For example, Figure 1.7b shows the further division of the 100 km grid square TL. Using this system, you can identify a 10 km square grid by giving first the two-letter code for the 100 km grid square followed by the easting and then the northing on the 10 km grid. So for example, in Figure 1.7c we can identify the 10 km grid square TL63.

On 1 : 50 000 OS maps, you can find the two grid letters together with two numbers (in brackets) e.g. TL (52) on the key or on the corner of the maps. At larger scales the grid has been further divided into 1 km intervals as shown in Figure 1.7c. As with the 10 km grid, we begin in the south-west corner and quote first the eastings then the northings. The example in Figure 1.7c identifies the 1 km grid square TL (52) 6432 where the (52) is just another way of identifying the 100 km grid square TL.

By estimating the distance between the 1 km grid lines, you can specify a position to within 100 metres, so for example the 100 metre grid reference for the point shown in Figure 1.7c is TL (52) 643327. If the map you are working with lies entirely within a 100 km grid square, the TL (52) prefix can be omitted.

---

of Geological Sciences (IGS) when it was formed in 1965. This was renamed **British Geological Survey** (BGS) in 1985. The geological maps which accompany this Course are selected from the large number which are published by the BGS, though some still bear the old IGS imprint.

1.4.3 MODERN GEOLOGICAL MAPS

The maps you will be using in this Course are printed in colour. If you examine the maps in the Home Kit you will find that they are indeed very colourful. What do the colours represent? These geological maps represent interpretations of where the solid rocks, the **bedrock**, would occur on the Earth's surface were all the soil, vegetation and buildings to be removed. The colours indicate what geologists call the **outcrop**, which is where particular rocks are presumed to be found below this superficial material. Figure 1.8 shows a geological map of an area in the Cotswolds; it tells us, for instance, that claystone forms the bedrock at Childswickham and Broadway but if we move south-eastwards towards Broadway Hill we can find limestone. How is such a geological map made?

The geologist in the field records the nature of the rock where it is visible at the surface, and these rock **exposures** are examined and characteristics such as rock composition, internal structure and fossil content noted. Using these records, different groups of rock can be distinguished and shown separately on the base map. Of course, although the whole of a map area is shown coloured according to the rocks present, this does not imply that at any point rocks of that particular type are actually visible at the surface, or that at some time a geologist has visited every exposure to record the rock type.

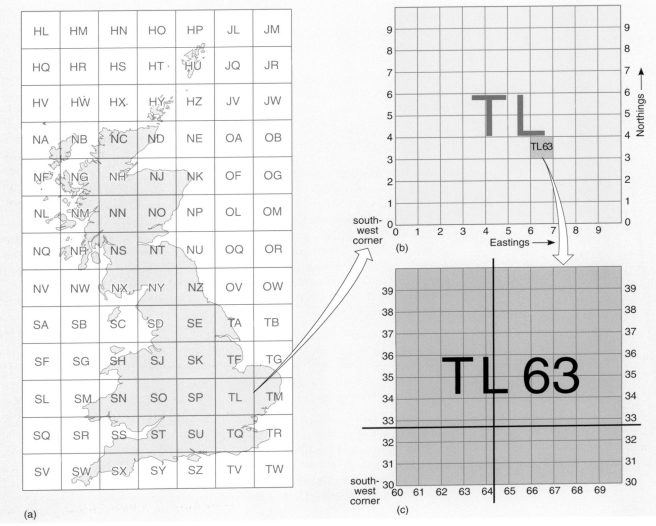

Figure 1.7 (a) The United Kingdom National Grid. The squares are 100×100 km and are represented by a two-letter code. (b) 100 km grid squares can be further subdivided into a 10-km-spaced grid. Using these 10 km grid lines, it is possible to specify any grid square by quoting first its easting, then its northing. (c) The 10 km grid is divided into a 1 km grid on 1 : 50 000 and larger-scale maps. Using the eastings and northings and by estimating the distance between the grid lines, it is possible to specify any point in the UK to within 100 metres.

In much of lowland Britain, the rocks are covered by soil and vegetation, or by urban construction or industrial debris. Deducing which rocks underlie these areas involves making use of basic geological principles and additional data such as the type of soil, the land's surface forms and information from boreholes to arrive at the best configuration that fits what can be seen. This brings us to an important consideration: even the process of actually making a map is interpretative, so no map should necessarily be regarded as being absolutely correct. That having been said, small-scale maps such as the BGS Ten Mile Map are unlikely to undergo radical revision tomorrow as a result of a new geological exposure because the map shows only a broad picture of the geology. However, a more detailed map such as the Cheddar Sheet you will use in Section 4 could be much more affected by new exposures revealed during the construction of a new road, for example.

The colours or shades on the map are separated from each other by sharp boundaries (Figure 1.8). These **geological boundaries** are lines showing where

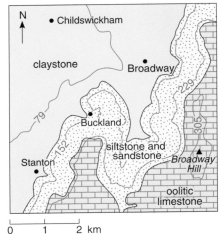

Figure 1.8 A geological map of the Broadway area in the Cotswolds. Contours in metres.

surfaces separating different kinds of rocks in the ground intersect the topographic surface and they allow us to determine the areal extent, sizes and shapes of the various bodies of rock. The geologist makes a map by recognizing in the field where one kind of rock ends and another begins and indicating that change by a line on the map. Nevertheless, there are always parts of the map where uncertainty exists about the nature of the bedrock, and it is important to realize that a good deal of interpretation is used in the map-making process and that not all maps are based on the same amount of information.

1.4.4 Scale in the S260 geological maps

The maps commonly used for geological purposes are based on Ordnance Survey topographic maps. Both OS and BGS maps are referred to as sheets because there is a lot of other information printed on the sheet around the edges of the actual map. Basic field mapping is carried out today in the UK on a scale of 1 : 10 000 (10 cm to 1 km, roughly 6 inches to 1 mile), though these maps are rarely printed. All the maps in your Home Kit are at a smaller scale and are a reduction and generalization from the 1 : 10 000 sheets on which the geologist works in the field. This means that there is progressive simplification and selection of data as smaller and smaller scale maps are prepared from the original field maps.

In Section 4, you will be working on the Cheddar Sheet which is an example of a 1 : 25 000 geological sheet (approximately 2.5 inches to 1 mile). As you will see when you use this map, field boundaries are marked and therefore it is easy to locate a particular feature, while at the same time the sheet covers a reasonably large area of countryside.

The scale which the Geological Survey has made most use of in Britain for published maps is 1 inch to 1 mile (1 : 63 360) (now largely being replaced by 1 : 50 000; 2 cm to 1 km). The Geological Survey decided to issue maps at this scale in the latter part of the 19th century since the 'one-inch' base maps were the largest scale then available from the Ordnance Survey. The 1 : 50 000 Moreton-in-Marsh Sheet is the example chosen for use on this Course. The BGS has also produced a new series of maps at the scale of 1 : 250 000 (approximately 1 inch to 4 miles). These maps not only cover a much larger area but they also include sea areas as well as land. Obviously, the methods used in geological mapping offshore are fundamentally different from those used on land. The example we have chosen is the Lake District Sheet which will be relevant throughout this Course and particularly useful at Summer School.

1.5 The ten mile map

 We come now to the map which perhaps should be of greatest assistance to you, the Ten Mile Map, so named because its scale of 1 : 625 000 is approximately 10 miles to 1 inch. The whole of Britain is covered in two sheets, north (N) and south (S), showing the geology of most of Northern Ireland as well as Britain. The base map shows roads, railways and rivers as well as towns. No contours are shown, although an indication of the relief is given by spot heights (in feet). You will become accustomed to using this map during your studies, and if you carry the map with you on journeys around the UK, you will find it a constant source of interest and information. The red grid overprinted on the map shows the boundaries of the larger scale one-inch and 1 : 50 000 geological sheets. The National Grid is overprinted in grey to allow easy reference to any locality by a series of digits (its National Grid reference) which may be found on OS maps at any scale.

The Geological Survey Ten Mile Map is divided into its two sheets along the 500 km Northing National Grid line. You will notice that the maps are described at the top of the key as '4th Edition'. The use of the word 'Solid' indicates that the rocks are shown as if there were no **superficial deposits** on top of the beds of solid rock. Superficial deposits include all unconsolidated material deposited during the Quaternary Ice Ages and all other materials deposited since, such as river gravels and alluvium, peat, and beach sands. A separate map showing only these superficial deposits is also published by BGS and is entitled Quaternary Geology.

Look first at the key to the South Sheet. The sedimentary rocks are set out in stratigraphic order, as they are on Plate 1.2. Eras and periods are marked down the right-hand side of the key. Each sedimentary rock has a number from the oldest (Precambrian, 60) at the bottom to the youngest (Pleistocene, 115) at the top but there are no radiometric ages given here. To understand the significance of the other terms in the key, we need to extend the previous discussion on the stratigraphic column. The individual coloured boxes in the key up to and including the Carboniferous Period represent time subdivisions of the periods. Thus Caradoc (70) represents a time interval during part of the Ordovician, and Namurian (81) a time interval during the Carboniferous. Neither name gives any clue to the types of rock to be found on the ground.

After the Carboniferous, the system used on the key changes and now represents different types of rock strata, rather than only periods of time, so it is called a **lithostratigraphic** (or rock stratigraphic) **column**. It can be divided into **rock units** for which there is a separate hierarchy of terms. The smallest unit (too small to show on the Ten Mile Map) is an individual **bed** of a particular rock type which is separated from those above and below by **bedding planes**. Beds may be grouped together into units called **members** and members can be grouped into **formations**. Formal formation or member names are denoted by upper case first letters, for example the London Clay (108). The divisions of the key to the Ten Mile Map for the Permian formations onwards are rock unit or formation names. If you look at the key for the Cretaceous Period, you will see that the units are mostly descriptive of rock types. Only the terms Hastings Beds and Gault are not descriptive (although Gault is actually an old name for mudstone).

The lithostratigraphic system is not used for the Carboniferous and earlier periods because well-defined rock units of a single type are not readily traced for long distances. In different areas, different rocks were generally being deposited at the same time so that strata of the same age are represented by different kinds of rock in different parts of the country.

The term used to describe the general characteristics (colour, texture, mineral composition, etc.) of a sedimentary rock is **lithology**. Rock units occurring at different localities, which may be of different lithologies, are said to correlate if they can be shown to be of the same age. Correlation of different rock outcrops between different areas of the country cannot therefore rely on recognizing a distinctive rock type. The complex mixed terminology in the key arose because until recently there were no generally agreed principles for assigning names to rock units for stratigraphic purposes. In the early 1970s, an International Code of Stratigraphic Classification was proposed which has since been adopted by most geologists. As the Code cannot operate retrospectively, many of the old names are still in use. Do not worry about these details of rock classification: the important points will become clearer as the Course proceeds.

The key to the igneous rocks is considerably simpler than that for sedimentary rocks. The igneous rocks are grouped into **intrusive** and **extrusive** (volcanic) types. The extrusive rocks are only broadly grouped according to the *time* during which they were erupted, and the coloured boxes are labelled according to the *type* of igneous rock represented. Thus, you cannot tell from this map whether an outcrop of basalt coloured pink and labelled 42 is of Silurian, Ordovician or Cambrian age. In other words, this part of the key is not a stratigraphic column in the strict sense. The intrusive igneous rocks are not divided by period at all, and are coloured and numbered by rock type. Finally, the metamorphic rocks are also labelled descriptively. There are only two categories, both of Precambrian age. The last items on the key are major **faults** and major **thrusts**. Faults and thrusts are more or less planar fractures along which movement of one side has occurred relative to the other and we will examine them in more detail in Sections 2.4, 3.4 and 4.5.

Turning now to the North Sheet of the Ten Mile Map, you will see that the key reflects the differences in geology between the north and the south of Britain. There are fewer sedimentary formations although they cover a similar time-span. (*Note:* There is an error in the key to the 3rd Edition of the Ten Mile Map (N): the oldest sediment, Torridonian (61), is incorrectly assigned to the Palaeozoic Era. It is in fact of Precambrian age, and should be shown as belonging to the Proterozoic Era. You should correct this part of the key on your map – the Palaeozoic bracket should extend only as far as the Cambrian.)

The range of igneous rock types is essentially the same as on the South Sheet but with the important difference that extrusive igneous rocks of the Tertiary Sub-Era are present in Scotland particularly on the islands of Skye NG (18) and Mull NM (17). There are no areas shown as Cambrian or Precambrian extrusive igneous rocks. This is because in this area rocks of these ages have been **metamorphosed** after the Cambrian Period and so are now classified as metamorphic rocks. The key to metamorphic rocks is much more detailed than on the South Sheet, reflecting the fact that the greater part of the Highlands of Scotland are composed chiefly of these rocks along with major igneous intrusions. The different boxes in this part of the key are again labelled descriptively but the rocks are divided into four categories. The first includes the metamorphosed igneous rocks referred to above. The other three (Dalradian, Moine and Lewisian Complex) are old formation names. They were defined before the International Code was established and do not fit into the modern classification. In any case these rocks are difficult to place in a stratigraphic succession, and the informal term **complex** (a 'complex' mixture of rock types) is best used for all three groups.

Several very prominent **fault lines** are shown on this sheet of the Ten Mile Map, and we shall examine these in Section 2.6. Another prominent feature of the North Sheet is the more or less regular pattern of red lines of basaltic rock which occur in the whole of western Scotland from northern Skye to the south-eastern boundary of the Dalradian Complex. These igneous rocks have been intruded into the rocks of the Highlands and Islands along steep or vertical cracks, and have solidified to form thin curtains of rock known as **dykes**. The dykes cut across geological boundaries existing at the time of the intrusive episode; if you look closely, you can see that they cut across the formation boundaries of the Dalradian Complex (e.g. at NS (26) 09). It follows that the dykes must be younger than the Dalradian rocks. We shall consider dykes in more detail in Section 2.6.

1.6 DIGITAL GEOLOGICAL MAPS

The increasing development of computers has lead to changes in both the production and the use of geological maps. The main technological development is the advent of the **Geographic Information System (GIS)** that can now be used for scientific investigations of geological data. For example, a

GIS might allow environmental geologists to identify areas where more porous rocks might pose pollution risks if a landfill site were located there. A GIS is a computer system capable of assembling, storing, manipulating and displaying data identified according to location. For geological information this might include solid geology, superficial geology, borehole information, seismic information or more detailed information on the solid geology such as chemical or fossil information.

GIS plays an increasingly important role in modern geology, in particular, where development and environmental issues are concerned. For example, suppose there is a proposal to convert a disused quarry to a landfill site. With a GIS, it is a simple matter to 'point' at a location, object, or area on the screen and retrieve recorded information about it from off-screen files. Using aerial or satellite images as a visual guide, it is possible to ask a GIS for information about the geology of the area or even about how close the proposed landfill site is to local rivers. This kind of analytical function enables conclusions to be drawn about the proposed site's environmental sensitivity. A GIS can also recognize and analyse the spatial relationships among mapped phenomena. Conditions of adjacency (what is next to what), containment (what is enclosed by what), and proximity (how close is something to something else) can all be determined with a GIS. By combining maps of geology, streams, slopes, land use etc., the GIS can be used to produce a new map that ranks proposed landfill sites according to their environmental impact. A critical feature of a GIS is its ability to produce graphics on the screen or on paper that convey the results of analyses.

1.7 SUMMARY OF SECTION 1

*   Uniformitarianism, the basic principle proposed by James Hutton, says that we can interpret the past through observation of the Earth today because processes have remained the same throughout geological time.

*   The principle of superposition states that sediments are deposited in layers, with the oldest at the bottom and successively younger layers on top.

*   The principle of faunal succession holds that because fossils succeed one another in order, rocks containing similar fossils are similar in age. This principle has allowed geologists to construct the geological time-scale, by which the relative ages of rocks can be measured.

*   Geological maps developed from the early ideas of William Smith who was the first to show the outcrop patterns of strata on a map of Britain.

*   Absolute ages of rocks can be determined by measuring the products of the radioactive decay of certain elements.

*   All geological maps have a key with a stratigraphic column of sedimentary rocks showing the oldest at the bottom and the youngest at the top. This interrelates the principles of superposition and faunal succession.

*   Igneous and metamorphic rocks are also shown on geological maps and in the keys to those maps.

*   Geological maps necessarily contain an element of interpretation in areas where the exposure of solid rocks is poor.

*   Geological maps contain information about the rock types, stratigraphic ages and structures of the rocks. Small-scale maps are less detailed than larger-scale maps.

*   Radiometric determination of geological ages has allowed dates to be assigned in millions of years to the geological time-scale.

1.8 OBJECTIVES FOR SECTION 1

Now you have completed this Section, you should be able to:

1.1   Locate a feature on any British map using the National Grid.

1.2   Be able to describe the age relationships of the stratigraphic units of Britain using the Ten Mile Map.

1.3   Be able to state the correct relative ages of rocks in terms of the eons, eras and periods on the stratigraphic column.

1.4   Be able to describe the main structural features of the geology of Britain.

Now try the following questions to test your understanding of Section 1.

Question 1.7   On your Ten Mile Map (S):

(a) What town is located at grid reference SY(30) 6990.

(b) On what sedimentary rock is it built?

(c) To what eon, era and period does this rock belong?

(d) Approximately how many million years ago were these sediments laid down?

Question 1.8   On your Ten Mile Map (S), locate the Isle of Man SC (24) 3080.

(a) What age are the youngest sedimentary rocks on the island?

(b) What age are the oldest sedimentary rocks on the island?

(c) List the sedimentary rocks present on the island in chronological order.

Question 1.9   Use the simplified geological map in Plate 1.2 to answer (a) and (b). For (b), you will need to refer to Figure 1.2.

(a) What are the approximate ages of the rocks of the northernmost tip of mainland Scotland and the southernmost tip of Ireland?

(b) What is the maximum difference in age between the rocks on which London and Dublin are built?

2 LANDSCAPES AND GEOLOGY

Throughout Section 2, you will need to refer frequently to your Ten Mile Maps.

2.1 THE DYNAMIC EARTH

The surface of the Earth's Moon is pockmarked with large numbers of impact craters as are the surfaces of many of the other planets and satellites of the Solar System. Yet when we look at the Earth, we see a very different landscape. Why is this? Why are there so few impact craters visible on the Earth? It is unlikely that the Earth has not been hit by asteroids or comets as much as has the Moon because they have occupied the same region of space for the past 4600 million years. One reason the Earth looks so different from other planetary bodies in the Solar System is its dynamism, i.e. the movement of its plates and its active volcanoes. Another is the fact that, unlike the other planets, the Earth possesses an atmosphere, liquid water and living organisms in abundance. These agents are involved in physically disintegrating and chemically decomposing the rocks and minerals of the Earth's surface, collectively known as **weathering**; less-weathered materials may then be eroded, in other words transported and deposited somewhere else. These processes of weathering and **erosion**, which together shape the landscape of a region, are so inconspicuous on a day-to-day time-scale that their importance is often underestimated, yet they are responsible for the majority of the landscapes that we see around us.

In the British Isles, as in most temperate lands, solid rock is not extensively exposed at the surface but is commonly covered by soil and superficial deposits. However, the form of a landscape can often tell the geologist a good deal about the underlying rocks because landscapes result from the processes of weathering and erosion on the one hand and deposition on the other. When preparing a geological map, therefore, the interpretation of landscapes may often provide important clues as to the underlying geology.

Rocks differ considerably in their resistance to erosion, so different rock types of differing lithologies may be subject to **differential erosion**. In general, among the sedimentary rocks, limestones and sandstones are relatively resistant to erosion and tend to form the higher ground with positive features such as cliffs and steep slopes, whereas less-resistant mudstones underlie the intervening valleys and commonly do not form outcrops. Igneous rocks tend to be very resistant to erosion. The relationship between landform and geology is best studied in the field and we will closely link work on extracts from Ordnance Survey maps with a 'virtual field trip'. We begin by looking at a landscape dominated by igneous rocks.

2.2 GRANITES IN SKYE

You have already seen on the Ten Mile Map that intrusive igneous rocks (mainly granites, shown in bright red) are abundantly exposed in Scotland although less common in England. Now look at these granite masses as they appear on the Ten Mile Map (N) on the Isle of Skye.

Plate 2.1 is a simplified geological map of the Isle of Skye. The Red Hills form an area of relatively high ground to the north of the Strathaird Peninsula and are composed of granite. The lower ground of the Strathaird Peninsula is underlain by much softer Jurassic sedimentary rocks capped in places by volcanic rocks. The sedimentary rocks weather more readily to form deeper soils and better farmland, while the volcanic lavas weather less readily and form the hills.

Figure 2.1 shows an important feature of granites. The rock contains large numbers of vertical and horizontal cracks or fractures called **joints.** Joints are planes of weakness; they are extremely common and are present in almost every exposure of rock. In igneous rocks, they develop primarily as a result of contraction (shrinking) on cooling. Granite is a plutonic rock, an igneous rock that was intruded beneath thousands of metres of overlying rocks. At depth, joints are kept closed by high pressure but, as the overlying rocks are removed by erosion, the gradual release of pressure allows the joints to open.

Figure 2.1   A photograph of an exposure of Shap granite from the Lake District showing vertical and horizontal jointing.

 You should now look at map extract 1 which is part of the 1 : 50 000 OS sheet covering the southern part of Skye. The south-west corner of the extract lies at NG (18) 5111 so you can locate the area on the Ten Mile Map. As the map extract lies entirely within the 100 km grid square NG (18), this prefix has been omitted from the grid references which follow. You should find the following places on the extract before reading further and you may find it helpful to mark these locations in pencil on the extract:

Broadford village (not labelled)	640237
Torrin	578210
Elgol	520140
Bla Bheinn	530218
Beinn na Caillich	602233
Camas na Sgianadin	625257

Using this map extract, work your way through the following questions.

Question 2.1

(a) Which areas of the map are occupied by the highest ground and which by the lowest?

(b) Follow the road north-east from Elgol: does the road have many steep gradients?

(c) Where are the flattest areas of the map? (*Clue:* examine the spacing of the contours.)

 ## Optional Activity

If you are unsure how to draw a topographic profile, refer to the Optional Activity in the Workbook.

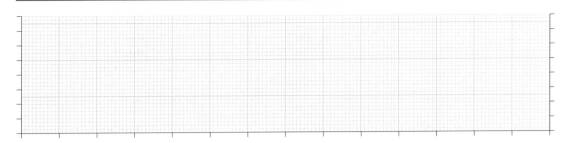

Figure 2.2   Blank graph paper for use with Question 2.2.

Question 2.2   Draw a line on the map from the promontory at Camas na Sgianadin (627261), through the summit of Beinn na Caillich to Torrin (580210) using the following points on the line: A (616248), B (601233), C (592223), D (582212). Using the blank graph paper in Figure 2.2, draw a topographic profile from D to A and mark points A–D onto it. The south-western end (D) is on the left-hand side of the profile. Use a scale of 1 cm = 500 m for the horizontal axis, and 1 cm = 250 m for the vertical axis.

Question 2.3   What is the average gradient between (a) A and B, (b) B and C, and (c) C and D?

You may be surprised to find that the gradient between B and C is fairly gentle despite the marked dip in the topographic profile at Coire Gorm. The two summits on your profile (Beinn na Caillich and Beinn Dearg Bheag) are both composed of granite. Coire Gorm is a small glacial depression formed from one of the streams running off the granite. However, the average gradients between A and B and C and D are much steeper.

❑ Look again at the geological map of Skye in Plate 2.1. With what geological features do the steep slopes between A and B and C and D coincide?

■ Edges of the granite.

The rocks around Broadford (Plate 2.1) are mainly Mesozoic sedimentary rocks; however, if you carefully examine the rocks along the line you drew, you will notice there is an outcrop of another igneous rock called gabbro to the north of Beinn na Caillich. The two changes in gradient between A and B most probably represent the boundaries between the granite and gabbro and between the gabbro and Mesozoic sedimentary rocks. The point at which the gradient changes is called a **break of slope**. Such marked breaks of slope are most commonly associated with a change in the underlying rock type. On the south of Beinn na Caillich, there is again a marked change in gradient coincident with the junction between the granite and the Jurassic sedimentary rocks around Torrin.

A typical intrusion, called a **pluton**, is shown in Figure 2.3, where you can see that the walls of the pluton slope downwards away from its centre at a steep angle. This is seen on the northern edge of the Red Hills where the granite crops out next to another igneous rock and there is a sharp steeply angled **contact** between the two (Plates 2.1 and 2.2). It is assumed that at the southern edge of the Red Hills, the granite slopes away very steeply under the Jurassic sedimentary rocks in a similar way. These rocks, into which the granite (or any igneous intrusion) has been intruded, are called the **country rocks** (see Figure 2.3).

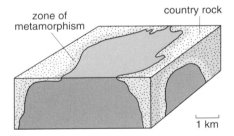

Figure 2.3   The three-dimensional shape of a pluton, showing the side walls and the zone of metamorphism that surrounds it.

❑ To convert the topographic profile you drew for Question 2.2 into a geological cross-section, draw a nearly vertical line at the break in slope between points C and D and at the two breaks in slope between points A and B. Label the rocks under Beinn na Caillich as granite, using a suitable symbol and colour, those to the south as sedimentary rocks and those to the north as first gabbro then sedimentary rocks.

■ Your cross-section should be similar to Figure A2.1 in the Answer section at the end of this Block.

Activity 2.1

Now please attempt Activity 2.1. This comprises two parts. First, you should view the video sequence *Geology and landforms* which looks at the topography of Skye. This video sequence also contains a short introduction to the DVD-based material for this Block and introduces Section 2.3. The second part involves a computer-based field trip (virtual field trip) to the same area.

The granites of Skye are a simple example of geological control of the landscape and we shall revisit this area in Activity 2.2. The large mass of erosion-resistant granites forms the high Red Hills, while the less-resistant sedimentary rocks to the south have been eroded more quickly and form lower ground (Plate 2.2). The picture for the rest of Northern Scotland, North Wales and the English Lake District is broadly similar, with erosion-resistant igneous or metamorphic rocks forming the high mountains and sedimentary rocks forming the low ground. In southern England, however, it is generally the relative resistance to erosion of individual sedimentary strata that is responsible for the topography. A good example of this is the area of the Dorset coast.

2.3 MESOZOIC STRATA ON THE DORSET COAST

On the Ten Mile Map (S) you will see that the distance from Abingdon (SU (41) 4797) to Camberley (SU (41) 8759) is about 50 km and between these two places sedimentary formations 99–109 crop out. The same formations are crossed in less than 5 km near the Dorset coast to the west of St. Albans Head at SY (30) 9675. The reason for the differences in the **width of outcrop** of the formations should become apparent as you read this Section.

2.3.1 UNIFORMLY DIPPING BEDS

The rock exposed in the quarry at Winspit in Dorset (Figure 2.4) is the famous Portland Stone, a limestone which has been widely used for building, e.g. St Paul's Cathedral. In addition to conventional quarrying, tunnels have been driven into the rock face to exploit a particular bed of rock approximately 2 m thick, without removing the overlying material. *Horizontal* sedimentary layering or bedding shows up well in this photograph, as do *vertical* joints. The quarrymen exploited these features when excavating the tunnels, with the roof and floor following bedding, and the sides following joints.

Suppose we are constructing a geological map of the formations along the Dorset coast and we come across an outcrop consisting of eroded upturned strata as shown in Figure 2.5. Because the layers were originally horizontal, i.e. deposited parallel to the Earth's surface, we know that they must have been disturbed in some way.

In order to describe the attitude or orientation of the layers, two features called the strike and the dip of the strata are used. Bedding and other geological layers or planar surfaces that are not horizontal are said to **dip**. Look again at Figure 2.5; we can see that the bedding-plane surface slopes toward the east. The compass direction towards which the plane slopes is called the **direction of dip** and can be visualized as the direction in which water would flow if poured onto the plane. The bedding-plane surface also forms an acute angle with the horizontal, called the **angle of dip**. We can also see in Figure 2.5 that the intersection of an imaginary horizontal plane with the bedding plane of one of the layers forms a straight line. This is the same as the line that would be formed by the edge of a pool of water bounded by the bedding plane. The compass

Figure 2.4   Quarry at Winspit, showing horizontal tunnels about 2 m high whose shape is controlled by the horizontal bedding planes and vertical joints.

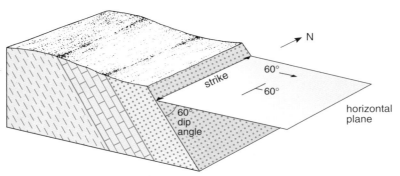

Figure 2.5   Block diagram of tilted strata illustrating the principles of strike and dip. The strike of a rock layer is the direction of the line formed by the intersection of the horizontal plane and the bedding plane. The dip of the layer is measured in a vertical plane perpendicular to the strike. The symbols show the two conventions used to indicate dipping beds on geological maps.

direction of that line is called the **strike** of the layer or bed. It gives the trend of the layer on the Earth's surface. Notice that the direction of dip and the strike are always at right angles to each other. In the field, the strike of a bed is recorded with a magnetic compass incorporating a device called a clinometer, which operates in the same way as a plumb line, to measure dip angles. To determine the strike, we could measure the direction of dip since the strike is 90° away from that direction. However, with most compass clinometers it is easier and more accurate to determine a horizontal direction and so measure the direction of strike directly. For example, as the beds in Figure 2.5 are dipping at 60° to the east, then the strike of the beds is 90° away, to the north. Both the strike and dip are depicted on geological maps because they enable us to grasp immediately the orientation of the layer with respect to the Earth's surface.

Box 2.1 Recording strike and dip in the field

When recording strike and dip, it is not necessary to record both the precise direction of strike and precise direction of dip because they are, by definition, at right angles to each other. For instance, strike and dip in Figure 2.5 can be recorded as 180/60 (in this notation, degree symbols are omitted), meaning a strike of 180° and dip of 60°. However, to indicate that dip is towards the east, rather than west, we need to add a letter to indicate approximate compass direction: in this case, we write 180/60E. We could equally well have written 000/60E since a strike of 000° is the same as a strike of 180°. Note: to avoid confusion between a dip and a strike measurement, the strike measurement should always be recorded as a three-digit figure.

❑ What would a value of 070/40N measured on a bedding plane signify in a notebook?

∎ It would indicate a strike of 070° and a bedding dip of 40° in a northerly direction.

In this case the dip is not exactly north. That does not matter because 070 is the precise reading and N merely serves to indicate the dip is toward the north, rather than toward the south. Moreover, since we know dip and strike are at right angles to each other, it follows that the precise dip and dip direction of the bed is 40° toward 340° (i.e. NNW).

Now look at map extract 2 which is at a scale of 1 : 50 000 and shows part of the Dorset coast west of Swanage (SZ (40) 0378) and also look at Figures 2.6 and 2.7. The area shows topographic and geological features commonly found in other parts of Britain underlain by sedimentary rocks, but here we can also see the strata exposed in the sea cliffs. You will probably find it useful to find and underline in pencil the following locations on the Dorset map extract, all of which lie within the 100 km grid square SY (30):

Winspit	977761
(Swanworth) Quarry	968783
St Albans Head	960754

❏   Is it reasonable to assume that the Portland Beds, shown in Figure 2.4, are horizontal all the way between Winspit and Swanworth Quarry?

■   Since there are no outcrops shown on the OS map, we will have to look at the relief of the area to help answer this question.

West of Winspit, the contours are widely spaced and the ground is clearly flat. This flat area is called a **plateau** and is the result of the ground surface being coincident with the flat bedding planes of the Portland Beds. The plateau is cut by cliffs at St Albans Head (Figure 2.6) where we can see that the plateau is underlain by a horizontal 'slab' of Portland Beds. The more complicated contour pattern to the north of Winspit has been produced by river valley erosion.

Now we shall look at the older rocks below the Portland Beds, for which you will need to find the following places

Renscombe Farm	965775
Kimmeridge	916796
Emmetts Hill	958763
Swyre Head	934783
Hounstout Cliff	950773

Look at the contours between Renscombe Farm and Kimmeridge. Several valleys run out to sea, with the ridges between cut off by the sea to form steep cliffs. Figure 2.7a is a view looking north-westwards from St Albans Head towards Hounstout Cliff. At Hounstout Cliff, the Portland Beds are the lighter-coloured rocks which form the top one-sixth of the cliff, beneath which there is a darker softer rock, the Kimmeridge Clay Formation.

Question 2.4   How can the soft Kimmeridge Clay Formation (a mudstone) form a steep cliff?

Figure 2.7b is a geological interpretation of the view shown in Figure 2.7a. At Emmetts Hill, the resistant Portland Beds form a vertical cliff, but the underlying Kimmeridge Clay forms a slope with a gradient between 30° and 60°. This slope has a series of steps, which were produced by a process known as landslipping. At the moment, the sea is not removing the material from the foot of the cliffs here as it is at Hounstout, but the cliffs are being eroded as huge blocks of Portland Beds slip seawards on the softer mudstones. **Landslips** of this kind are common in both coastal and inland localities of Britain where hard rock units overlie soft mudstones.

From the view shown in Figure 2.7a, it is clear that the cliffs are cutting across a landscape of valleys and hills. It is reasonable to assume that at one time the Portland Beds were a continuous flat sheet between Emmetts Hill and Hounstout Cliff, and that Swyre Head to the west of Hounstout Cliff is also capped by these beds.

Now find:

Smedmore House	925788
Orchard Hill Farm	941796

Figure 2.6 Aerial view of the limestone plateau and cliffs in the vicinity of St. Albans Head. Note that there has been some quarrying at St. Albans Head, which accounts for the large fallen blocks of rock at the foot of the cliffs. Elsewhere, fallen material has accumulated by natural processes.

Swyre
Head

Hounstout
Cliff

Emmetts
Hill

Figure 2.7 (a) Photograph looking north-westwards from north of St. Albans Head. (b) Sketch of the view in (a) to show the cliffs below Emmetts Hill, indicating the origin of the stepped cliff profile by landslips.

(a)

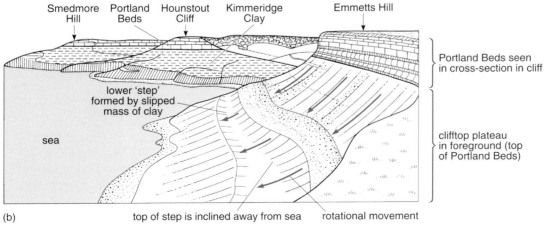

Smedmore   Portland   Hounstout   Kimmeridge   Emmetts Hill
Hill   Beds   Cliff   Clay

Portland Beds seen in cross-section in cliff

lower 'step' formed by slipped mass of clay

sea

clifftop plateau in foreground (top of Portland Beds)

(b)   top of step is inclined away from sea   rotational movement

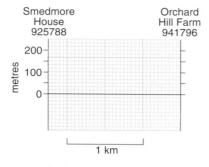

Figure 2.8   Graph paper for use with Question 2.5.

Question 2.5   Construct a topographic profile between these two points using the graph paper in Figure 2.8. Exaggerate the vertical scale by a factor of about 3 to emphasize the changes in gradient.

If you were to take a walk on a north-east line from Smedmore House to Orchard Hill Farm, you would first climb a moderately steep slope which would become very steep as you got to the top of the hill above Smedmore House, as indicated by the closer spacing of the contours on the map. Then going down to Orchard Hill Farm, the gradient would be quite gentle to the north-east. From the map extract, you can see Swyre Head is the south-eastward continuation of the hill for which you have drawn a topographic profile, and is capped by Portland Beds (see Figure A2.2 in the Answer section at the end of this Block). The very gentle slopes to the north-east in both cases must be equivalent to the plateau north of St Albans Head but now tilted a little.

The gentle slope which roughly parallels the inclination of the strata is termed the **dip slope**, whereas the steep slope facing away from the direction in which the Portland Beds are inclined or dipping is termed the **scarp slope**. Dip and scarp slopes are very common in areas of dipping strata of differing resistance to erosion, and provide important information on the disposition of the rocks.

The crest of the slope between Smedmore House and Swyre Head is fairly straight, but to the east of Swyre Head it becomes sinuous where the slope is cut by several valleys. Figure 2.9 is a topographic profile and geological cross-section across this area, from Smedmore House to Renscombe Farm, showing the boundary between the Portland Beds and the Kimmeridge Clay. From this Figure, you can see quite clearly that the heights of the hills decrease towards the east, reflecting the slight dip of the Portland Beds in that direction. So the Portland Beds once extended over the whole area, as a continuous and very gently inclined sheet which has since been eroded by river valleys.

Figure 2.9   Topographic profile from Smedmore House to Renscombe Farm showing the base of the Portland Beds. × 3 vertical exaggeration.

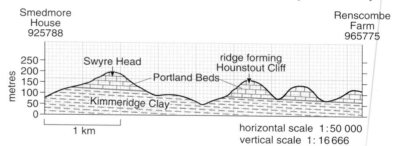

Now follow the scarp of the Portland Beds westwards from Smedmore House. Where it reaches the coast, it forms spectacular cliffs (over 150 m high) at Gad Cliff (886796). The Portland Beds here are tilted steeply to the north at an angle of 30° to the horizontal, that is, they dip at 30° in a northerly direction. There is a progressive increase in the dip of the Portland Beds going to the west from the horizontal strata at St Albans Head (Figure 2.6) to the gentle north-easterly dip at Smedmore Hill (Figure 2.7b) to the 30° dip at Gad Cliff.

Question 2.6   Examine the map extract around 1.5 km to the north of the scarp formed by the Portland Beds at Gad Cliff. How would you describe the topographic feature indicated by the contours? Can you suggest a possible geological cause for this feature?

2.3.2   TOPOGRAPHY AND GEOLOGY IN DORSET

Now find:

Worbarrow Tout	869796
Worbarrow Bay	865800
Rings Hill	864807
Whiteway Hill	875812

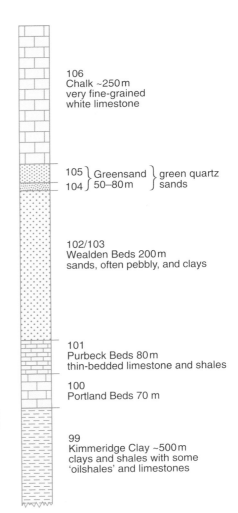

West   Rings Hill   East

Worbarrow Bay

Worbarrow Tout

sea-level

106 Chalk   105   104 Greensand   102/103 Wealden Beds   101 Purbeck Beds   100 Portland Beds

- The ridge formed by the Portland Beds runs into the sea at Worbarrow Tout. Imagine walking round the beach, from Worbarrow Tout to Rings Hill. Would you see rocks older or younger than the Portland Beds exposed in the cliffs as you walk northwards?

■ Younger rocks, because the beds dip to the north.

Figure 2.10 is a section across the bay; if you had difficulty in understanding why you would cross on to younger rocks as you went north, the reason should be clear from this diagram. Try positioning the diagram so that the strata are in the horizontal position in which they were laid down. Figure 2.10 also shows the beds which are responsible for the scarp you examined to the north of Gad Cliff; the scarp is formed of the steeply dipping Chalk.

Going north-west across Worbarrow Bay would enable you to work out the succession of rock formations in this area and these can be shown as a key on the map in which the sedimentary rock formations are shown in their correct chronological order (oldest at the base, youngest at the top).

As you saw in Section 1, there is a simple stratigraphic column (key) on the Ten Mile Map, but no indication of rock thicknesses is given. However, on larger-scale geological maps, the stratigraphic column is most commonly drawn to scale to indicate the thicknesses of the individual formations present in the area. An example of a simplified stratigraphic column drawn to scale from the BGS 1 : 50 000 geological map of the Worbarrow area is given in Figure 2.11. You need not remember the details of this particular column, but you should note that columns drawn to scale are another facet of the greater geological detail shown on larger-scale geological maps. Compiling a stratigraphic column is not always such an easy task, especially when there are few outcrops of solid rock to be seen, as is normally the case in inland Britain.

The block diagram in Figure 2.12 (overleaf) summarizes the relationship between topography and geology in this part of Dorset. The plateau area and ridges are underlain by erosion-resistant limestone rocks, the Portland Beds and Chalk, whilst the valleys have formed over softer mudstones and sands, which have been readily excavated by erosion to form bays, such as Worbarrow Bay. East of Kimmeridge the dip slope of the Portland Ridge is shallow, reflecting the gentle dip of the underlying strata. But to the west of Kimmeridge the dip slope of the Portland Beds is much steeper, reflecting the steeper dip of the strata there.

2.3.3 FOLDING AND WIDTH OF OUTCROP ON MAPS

Examine the outcrop pattern of the Worbarrow area on the Ten Mile Map (S), SY (30) 870799. You can see that the outcrop pattern trends east–west. As we saw in Section 2.3.1, the direction of dip is at 90° to the strike.

- How could you decide whether the strata in the Worbarrow area are dipping to the north or the south?

■ You have seen (Section 2.3.2) that going round the beach at Worbarrow Bay (Figure 2.10) there are progressively younger beds and they are dipping to the north.

Figure 2.10 Cross-section showing the geology and topography to the east side of Worbarrow Bay. (Note that the section follows the curve of the eastern side of the bay, so the relative dip of the Chalk is greater than shown here.) The names used for the strata are those used on the Ten Mile Map and the vertical scale is exaggerated.

106
Chalk ~250 m
very fine-grained
white limestone

105 } Greensand } green quartz
104 } 50–80 m } sands

102/103
Wealden Beds 200 m
sands, often pebbly, and clays

101
Purbeck Beds 80 m
thin-bedded limestone and shales

100
Portland Beds 70 m

99
Kimmeridge Clay ~500 m
clays and shales with some
'oilshales' and limestones

Figure 2.11 Stratigraphic column for the Worbarrow/Kimmeridge area. The numbers used for the strata are as on the Ten Mile Map.

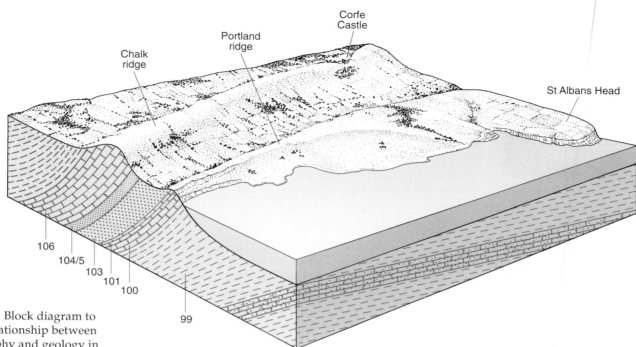

Figure 2.12 Block diagram to show the relationship between the topography and geology in the Purbeck area.

Remember (Section 2.3) that you found that the distance from Abingdon (SU (41) 4797) to Camberley (SU (41) 8759) is about 50 km and between these two places sedimentary formations 99–109 occur. The same formations can be crossed in less than 5 km near the Dorset coast to the west of St Albans Head at SY (30) 9675. Now look at how the width of outcrop of the Purbeck Beds (101) increases going eastwards from Worbarrow Bay: the Purbeck Beds are dipping quite steeply northwards at Worbarrow Bay but further east they become almost horizontal, like the Portland Beds (100). Provided that there are *no major changes in the thickness* of a rock formation, then as the dip decreases, the outcrop width of a rock formation *increases*. This is a general rule which is particularly useful when you use the Ten Mile Map where dip and strike symbols are not shown. You will get an opportunity to examine the relationship between dip and width of outcrop for yourself by means of the interactive models in your Block 1 CD-ROM, as part of Activity 4.2.

 Question 2.7 Look at the outcrop pattern of formations 104–109 at the east and west ends of the Isle of Wight on the Ten Mile Map (SZ (40) 38 and SZ (40) 68). In which direction are the strata dipping?

 Question 2.8 Examine the outcrop width of the Chalk (106) on the Isle of Wight, from the Needles in the west (SZ (40) 3085) to south-west of Newport (SZ (40) 5088) in the middle of the island. Assuming that the topography of the Isle of Wight is flat, explain why the dip of the strata must be very steep at the Needles and rather shallow south-west of Newport.

Earlier (Section 2.3.2), we said that the dip of the Portland Beds increases along the scarp westwards from St Albans Head (compare Figures 2.9 and A2.2). Such variation of dip is often evidence that the beds have been involved in deformation.

Folds are bends or kinks in sedimentary strata. A series of folds similar to those that are seen further down the Dorset coast at Lulworth Cove (Plate 2.3) are depicted in Figure 2.13 together with some of the terms used to describe folds. You will examine folds in greater detail in Block 3; here, we will introduce a few of the descriptive terms necessary for the study of folds as they relate to geological maps. A fold may be divided more or less evenly along its length to produce two parts called the **fold limbs**. Two types of folds, an **anticline** and a

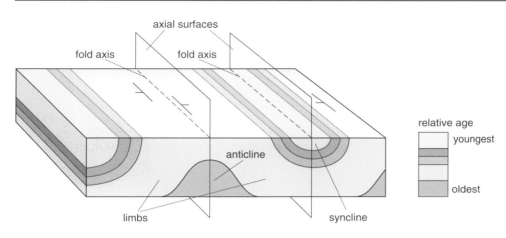

Figure 2.13 Anticlines and synclines in eroded strata. In an anticline, the limbs dip away from the axial plane. On a map, progressively older strata are found toward the axis of the fold. These relationships are reversed in the syncline; the limbs dip toward the axial surface and on a map progressively younger strata are found toward the fold axis.

syncline, are shown in Figure 2.13 and an example of a large syncline is shown in Plate 2.4. The limbs of an anticline dip away from the centre of the fold, giving the fold an arch-like appearance. On a map, therefore, the strata of the limbs will dip away from the centre of the fold. Because of this orientation, erosion will expose the older strata at the centre of the anticline and progressively younger strata on the limbs. The relationships are reversed for the syncline where the limbs dip toward the centre giving the fold a trough-like appearance, so that younger strata are exposed at the centre of the fold and the strata dip toward the centre of the fold on a map.

When strata are deformed into folds and then eroded, a single rock formation will have a complex outcrop pattern on a geological map. Now we shall look on the Ten Mile Map for evidence of folded beds.

❑ You already know that the Portland Beds and the Chalk dip towards the north in the Worbarrow area. Which way is the Chalk (106) dipping at Blandford Forum (ST (31) 8806)?

■ To the south-east, because younger strata (107, 108) occur to the south-east of the Chalk (106) and older strata (105, 104) to the north-west. Therefore, there must be a fold (a syncline) in the beds between Worbarrow and Blandford Forum.

In fact, this is one end of a large synclinal structure called the Hampshire Basin, whose edges can be roughly traced out on the map by following the outcrop of the Chalk (106). The northern side of the Basin runs from Dorchester (SY (30) 6990) to Salisbury (SU (41) 1530) and then south-eastwards towards Brighton (TQ (51) 3205); the southern side runs from Dorchester through Studland (SZ (40) 0482) and eastwards through the centre of the Isle of Wight.

We shall return to the patterns of folded strata on the Ten Mile Map when we look at the overall structure of Britain in Section 2.6 and take a more detailed look at folded strata on the Cheddar Sheet in Section 4.4.

2.4 CARBONIFEROUS STRATA IN THE NORTH PENNINES

Sedimentary rocks of Carboniferous age are widespread in Britain, reflecting the fact that during the Carboniferous much of Britain was covered by sea (Plate 1.1) that resulted in the deposition of limestones and other sedimentary rocks. Look at the key to the Ten Mile Map (S) and find the Carboniferous Limestone Series (80); it is pale blue. You will see that the Peak District area of Derbyshire (SK (43)) and the Yorkshire Dales centred on Malham (SD (34) 9061) are both underlain by these rocks. You should find SD (34) 77, the area to the north-east of Ingleton on the edge of the Dales. Most of this grid square is Carboniferous Limestone Series.

2.4.1 LANDFORMS OF THE CARBONIFEROUS LIMESTONE SERIES

 You will now need map extract 3 which is part of the 1 : 25 000 OS map of the Three Peaks and has its south-western corner at SD (34) 6972. It shows Ingleton and Ingleborough Hill. The rocks of this area are generally very well exposed and there are lots of clues to the underlying geology which you will be able to find for yourself as you study the map extract. Find the following places on the extract. (All these localities are in 100 km grid square SD (34) so this prefix has been omitted.)

Ingleborough Hill	741746	Twistleton Scar End	7075
Gaping Gill Hole	751727	Ingleton	6973
Meregill Hole	741757	Chapel-le-Dale valley	738772
Raven Scar	721746 to 734760	(Ingleton to Chapel-le-Dale)	

Look at the contours and rock features shown on the map extract and try the following question. (Answer each part in one or two sentences.)

Question 2.9

(a) What is the contour interval on this map, and how frequent are the heavier contour lines?

(b) Where is the lowest land?

(c) Where is the highest land, and how high is it?

(d) Describe how the rock features shown as rock outcrops on the map relate to the relief.

We shall now investigate how these features are related to the underlying geology. You know from your examination of the Ten Mile Map (S) that most of the area of the map extract is underlain by rocks of the Carboniferous Limestone Series. At the base are limestones formed from the skeletons of marine organisms. The beds are typically a few metres thick. The cliffs (scars) depicted on the map are usually formed by beds of limestone that are particularly resistant to erosion, the top of the cliff being marked by a bedding plane. Erosion of softer beds means that the harder ones stand out as features. The eroded rock collects at the foot of the cliffs as a steep slope of angular fragments, known as a scree deposit (Figure 2.14).

The scars shown on the map can be used to deduce that the limestone is still approximately in its original horizontal orientation, that is, it has little or no dip. Look at the cliffs at Twistleton Scar End.

 Question 2.10 What is the relationship between these cliffs and the contour lines?

 Question 2.11 Use the blank graph paper in Figure 2.15 (margin on opposite page) to draw a topographic profile from X (690734) to Y (750748) using only the thick 100 foot contours. (Use a vertical scale of 1 cm to 500 feet.) The horizontal (map) scale is 1 : 25 000, so 1 cm represents 250 m or 820 feet. Therefore, your section will have about a × 1.6 vertical exaggeration. On your topographic profiles to the south-west of Ingleborough Hill, mark three points:

P, where the line of profile crosses the 2300 foot contour;

Q, where the line of profile crosses the 1500 foot contour;

R, where the line of profile crosses the 1300 foot contour.

Calculate the average gradients (a) between P and Q, and (b) between Q and R.

The change in gradient along the topographic profile you constructed in Question 2.11 is quite dramatic: there is a plateau between points R and Q, then a steep slope up to point P. Such marked breaks of slope are generally associated with a change in the underlying geology.

Figure 2.14 Horizontal beds of Carboniferous Limestone forming cliffs, with scree deposits, at Raven Scar, Chapel-le-Dale, with Ingleborough Hill behind. The higher part of Ingleborough Hill above the limestone scars is formed chiefly of shales protected from erosion by a few resistant sandstone layers

Figure 2.15 (margin) Blank graph paper for use with Question 2.11.

If you examine the map extract you will notice that there are no streams on the limestone plateau areas above the cliffs. Although limestone is only sparingly soluble in pure water, when water passes through peat or soil its acidity can increase enabling it to dissolve limestone. As rainwater passes down the vertical joints and along bedding planes, solution of the limestone opens up fissures and this in turn allows water to pass more rapidly through the rock. On the surface, solution along joints results in the formation of features known as **limestone pavements** (Figure 2.16). Limestone pavements have developed in many areas of the Yorkshire Dales, especially in thick beds of limestones near the base of the Carboniferous Limestone Series.

Figure 2.16 Part of a limestone pavement in the Ingleborough area.

Ingleborough Hill is capped by the Millstone Grit and the upper slopes of Ingleborough Hill, above the 1500 foot contour, have surface streams. The rocks directly overlying the Carboniferous Limestone Series are a sequence of shales with some thin sandstones. The shales do not allow water through them and are thus impermeable so that water is carried

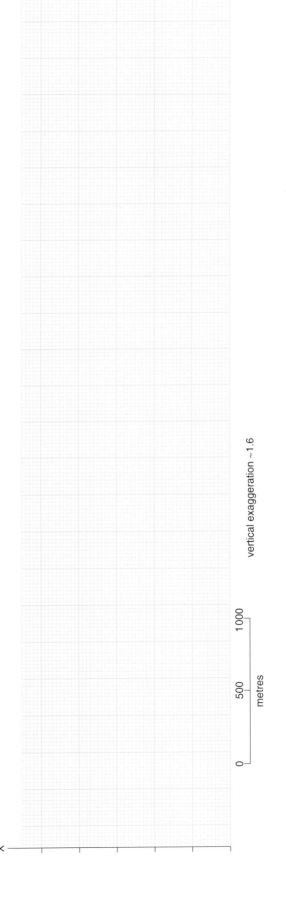

750748
Y

vertical exaggeration ~1.6

1000

500

metres

0

690734
X

on the surface in streams. When the streams reach the limestone plateau, they pass underground via potholes, which are commonly found at the junction of two or more major joints in the limestone (Figure 2.18). Gaping Gill Hole is a famous pothole at the entrance to a system of underground caverns and tunnels produced by solution along joints and bedding planes. The main cavern beneath Gaping Gill Hole is 110 m (365 ft) high. Enlargement of such caverns may be followed by collapse of the overlying rocks causing the formation of gorges at the surface, such as Cheddar Gorge (Section 4).

Question 2.12   Between what heights on the map extract are most potholes to be found?

On the topographic profile from Question 2.11, if you draw a horizontal line from Q beneath Ingleborough Hill, you can mark the strata above this to indicate the shales with thin limestones and sandstones (shales are usually represented by dashed horizontal lines and sandstones by dots).

Most stratigraphic units have parallel tops and bases, thereby making them of constant thickness, at least over distances which are small compared with the total geographical extent of the unit. This is of great practical importance when constructing geological cross-sections, so the tops and bases of strata that you draw in vertical cross-sections such as that for Question 2.11 are normally drawn parallel with each other, in this case horizontally. Only if you can find information on the map to tell you that the units are not of constant thickness would you draw them otherwise.

2.4.2   UNCONFORMITIES, INLIERS AND OUTLIERS

You saw in Section 1.2 that the Principle of Superposition determines the order in which sedimentary rocks are deposited; however, it does not imply that all layers have been deposited continuously over time. Indeed, if the geological record of a particular area were thought of as a pile of manuscript pages for a book, we would find that several chapters were absent as if they had not been written or had been removed. In the spring of 1788, James Hutton saw horizontally bedded red sandstones lying directly on vertical beds of grey–green rocks in the cliffs of Siccar Point in south-east Scotland (Plate 2.5). He was the first to infer that a sizeable section of the geological record was missing and that this kind of junction represents a long time interval between the two rock units during which the lower one was deformed and then partially eroded. A gap such as this in the rock record is referred to as an **unconformity** and can be identified by an erosional surface between rocks of markedly different ages. It indicates that there was a break in the deposition of sediments and a period of erosion between the rocks above and below. However, although an unconformity is evidence of a major gap in the geological record, we have no way of knowing how long a period of time that gap represents – unless we can get accurate ages for the rocks above and below the unconformity.

Two features that commonly arise in association with unconformities, though they can be produced in other ways as well, are inliers and outliers. An area of older rocks that appear to be completely surrounded by younger rocks is called an **inlier**. The reverse situation, of younger rocks surrounded by older, is an **outlier**. Both effects can also be produced by differential erosion (Section 3.3.5), but most inliers and outliers visible on the Ten Mile Map are produced where the junction between the younger and older rocks is an unconformity or a set of faults, or a combination of the two. Inliers are in effect 'windows' through the overlying rocks to what is below. Outliers are remnants of the younger rocks that once were more extensive.

Find the following localities on map extract 3:

Skirwith Cave   709738
White Scar Cave   713745
Ingleton Granite Quarries   718753
God's Bridge   733764
Long Chimney   698747
Thornton Force Waterfall   695754

The map extract shows a number of springs in Chapel-le-Dale valley, for example, in square 7175 there are ten springs marked, many of which are the source of small streams. You will notice that they all occur roughly in a line between 750 and 800 feet on the north side of the valley, known as a **spring line**.

❏   What does this indicate about the strata below this height?

■   This would indicate that below this height the strata must have very low permeability so that water cannot pass through them, in contrast to the permeable limestone immediately above, which has no surface drainage because water has passed directly into the joints.

Look at this area on the Ten Mile Map (S) where there is a 'finger' of (pale mauve) rock 'pointing' north-east from Ingleton.

❏   What is the age of this rock: is it older or younger than the Carboniferous Limestone Series? What term would you use to describe this outcrop pattern.

■   It is Ordovician, and it forms the floor of the valley of Chapel-le-Dale. This small area of Ordovician strata is totally surrounded by younger rocks of the Carboniferous Limestone Series and is an inlier – a 'window' through to the rock surface on which the Carboniferous strata must have been laid down.

In this area, Carboniferous strata lie directly on steeply dipping Ordovician rocks; in other words, strata from two whole periods, the Silurian and the Devonian, are not found here, representing a time gap of 100 million years. Either Silurian and Devonian rocks were never laid down in this area or they have been completely removed by erosion. With a gap such as this, the older rocks have been folded and thus dip at a steeper angle than the strata above. This is referred to as an **angular unconformity**.

❏   The lowest Carboniferous beds are made up of pebbles and boulders that have been cemented together to form a rock called a **conglomerate**. If you were able to date a pebble from the conglomerate, what age might you expect to get?

■   The pebbles must be fragments of rocks older than the Carboniferous. If no deposition took place in this area during the Silurian and Devonian, the pebbles are most likely to be composed of Ordovician rocks since they probably would have been eroded locally from the Ordovician rocks exposed in Carboniferous times.

The conglomerate thus lies directly on the irregular eroded surface of the older Ordovician strata (Figure 2.17). The unconformity is exposed at Skirwith Cave, which is on the topographic profile you drew for Question 2.11. If you draw a horizontal line on your section from Skirwith Cave beneath Ingleborough Hill, this will represent the plane of the unconformity. Mark the rocks above the unconformity with a 'brickwall' symbol to represent limestone (as in Figure 2.18).

On the south-east side of Chapel-le-Dale valley, the unconformity is also exposed at White Scar Cave and at the spring near the Old Quarry (720753). Join these points on the map extract through the intervening springs and you are now beginning to map the line where the unconformity may crop out on this side of the valley. Similarly, on the north-west side of the valley, the spring line in square 7175 can be drawn in to extend the unconformity south-westwards to Long Chimney and crossing the valley floor near God's Bridge. The spring line in Chapel-le-Dale is therefore a very good indicator of the outcrop of the unconformity.

Figure 2.17 The unconformity at the base of the Carboniferous limestone at Thornton Force, showing the horizontal limestone, the pebble conglomerate and the near-vertical Ordovician rocks (locally known as 'Ingletonian') beneath.

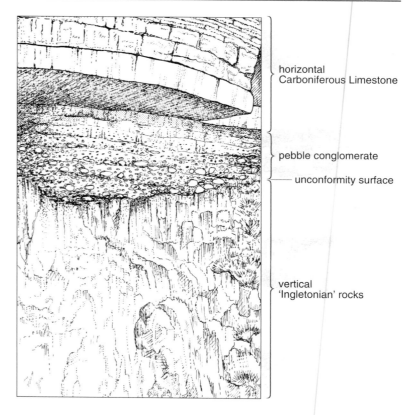

The Ordovician rocks in Chapel-le-Dale consist of several different sedimentary rocks – some were laid down as coarse-grained sands and others as fine-grained muds. All have since been subjected to pressure and heat (metamorphosed) and now comprise a steeply dipping sequence of impermeable rocks. Water percolating down joints in the Carboniferous Limestone is forced to move laterally when it meets the impermeable Ordovician rocks and thus appears as springs where the unconformity is at the surface (Figure 2.18).

Figure 2.18 Cross-section of Carboniferous Limestone showing how the potholes at the surface may be connected by tunnels and caverns to springs at an unconformity.

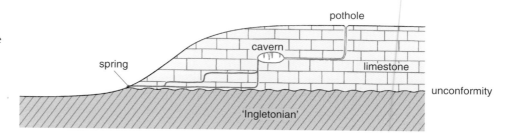

In the 1790s, Hutton's colleague John Playfair visited Chapel-le-Dale and described the relationship between the horizontal limestone, the conglomerate and the steeply dipping underlying rocks, both at White Scar Cave and at Thornton Force Waterfall (Figure 2.19 opposite page). Playfair recognized that this unconformity represents a huge time gap like the one Hutton described at Siccar Point in Scotland.

2.4.3 FAULTS

Folds and faults are structures produced by the deformation of rocks. We saw in Section 2.3.3 that folds are a response to deformation where the layers of rock are deformed without breaking so that the layers are curved but continuous. Faults represent a different kind of response of rocks to deformation, one in which the layers of rock fracture and move relative to each other.

Carboniferous Limestone

plane of unconformity

'Ingletonian'

Figure 2.19   Thornton Force Waterfall. The stream falls from the horizontally bedded Carboniferous Limestone into a valley cut in vertically dipping Ordovician rocks.

For most purposes, geological faults are grouped according to the nature of the relative movement that has occurred. The anatomy of a fault is shown in Figure 2.20. The surface along which movement has taken place is called the **fault plane**. There are two kinds of faults depending on the nature of the movement that has taken place. If the fault movement has a vertical component, along the dip of the fault plane (Figure 2.20a,b) the fault is referred to as **dip–slip**; the example shown in Figure 2.20a is a dip–slip **normal fault** and the example in Figure 2.21b is a reverse dip–slip fault. The extent of the vertical movement is called the **throw** of the fault. If the displacement has been horizontal, however, parallel to the strike of the fault plane, the fault is referred to as a **strike–slip** fault (Figure 2.20c). You should realize, however, that faults can have both vertical and horizontal components to their movement so that not all faults are pure dip–slip or pure strike–slip.

Look again at Ingleton village on the Ten Mile Map. Find where Barren Red sandstones (84) crop out at the surface.

❏   Are these rocks older or younger than the Carboniferous Limestone Series?

■   They are younger.

❏   How could these younger rocks have reached this much lower position?

■   An interpretation is that there is a fault which has brought the Barren Red sandstones against the Carboniferous Limestone Series (Figure 2.21). This fault (the Craven Fault) has been calculated to have a throw of about 2000 m.

In fact, movement has occurred along several parallel faults which are quite close together (within a kilometre) but these can be shown individually only on a larger-scale map than the Ten Mile Map. This series of faults is known as the Craven Fault System though they are shown as a single fault on the Ten Mile Map. If you compare Figure 2.21 with Figure 2.20, you will see that the Craven

fault plane

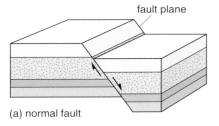

(a) normal fault

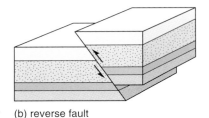

(b) reverse fault

(c) strike-slip fault

Figure 2.20   Some common types of faults and their associated terminology. Relative motion of the opposing blocks on either side of the fault plane determines whether a fault is classified as a dip–slip fault, either normal or reverse, or a strike–slip fault. A thrust fault is a low angle reverse dip–slip fault.

Figure 2.21 Cross-section through the Craven Fault System, represented as a single fault, showing the strata faulted down to the south-west.

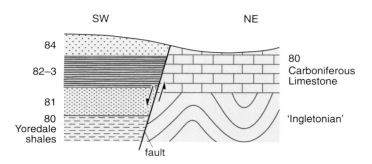

Fault System at Ingleton has a dip–slip normal fault movement. You will meet the other types of fault in Sections 3 and 4.

By examining the Ten Mile Map you will see that the Craven Fault System runs north-west to south-east, passing to the north-east of Ingleton and extends south-east to Settle (SD (34) 8263). In fact, this major fault is shown as a geological boundary between formations 81, 82–3, 84 and 89 to the south-west and formation 80 to the north-east.

It is difficult to see from the Ten Mile Map where this fault would be found on the OS map extract. From field evidence, the major fracture crosses the line of the profile you drew for Question 2.11 at 696735 (about 2.4 cm from X). Draw an almost vertical line (dipping at about 80° to the south-west) at this point on your profile to represent the fault; you can then mark with an arrow the left-hand (south-west) side as having moved downwards with respect to the north-eastern side (compare Figure 2.20). You can now show the rocks below the unconformity to the east of the fault as vertical Ordovician (Ingletonian) strata and the younger Carboniferous Barren Red sandstones to the west of the fault (use dots to represent the sandstones). The profile is now a simplified geological section across this area.

By drawing several cross-sections, a block model can be constructed to summarize all the topographic and geological features that have been discussed. Compare your cross-section with the right-hand face of Figure 2.22. This should help you visualize the three-dimensional aspects of the strata and their relationship with the topography.

Look again on the Ten Mile Map at the country to the east of Ingleton, where it is possible to infer that the basic geological structure as shown on your section is typical of that of most of the Yorkshire Dales. The lower ground is formed of the Carboniferous Limestone Series (80) with the higher peaks such as Ingleborough Hill capped by the overlying Millstone Grit Series (81). Whernside (SD (34) 7582) and Buckden Pike (SD (34) 9778) are other examples. In a few places, streams have cut down through the unconformity at the base of the Carboniferous Limestone Series to expose much older Ordovician rocks, for example, south of Horton in Ribblesdale (SD (34) 8172).

2.5 MAKING A GEOLOGICAL MAP

Activity 2.2

Before beginning detailed study of the larger-scale geological maps (Sections 3 and 4), please do Activity 2.2 in which we shall try to summarize many of the main points we have covered so far by producing a geological map of an area of the Strathaird Peninsula on Skye. The area was first visited as part of Activity 2.1 and you may wish to refer to Plate 2.1 to locate the Strathaird Peninsula.

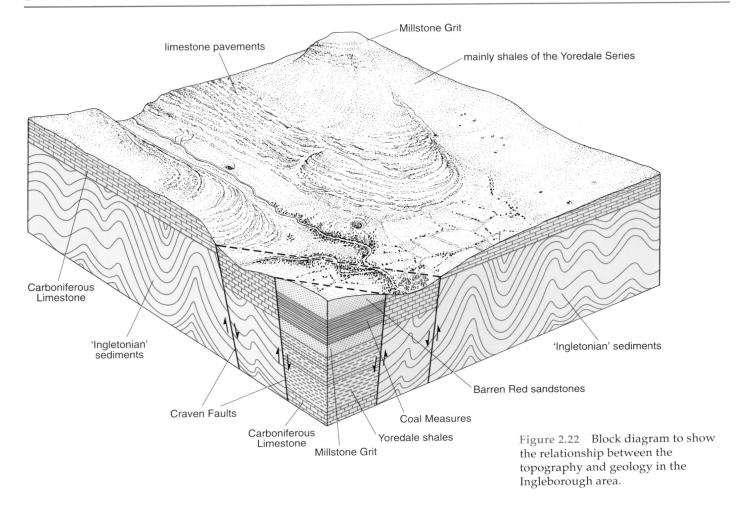

Figure 2.22 Block diagram to show the relationship between the topography and geology in the Ingleborough area.

2.6 SOME GEOLOGICAL FEATURES OF BRITAIN

Before you begin the more detailed study of larger-scale maps in Sections 3 and 4, we shall try to summarize some of the main geological features of Britain which can be seen from the outcrop patterns on the Ten Mile Map and the Geological Map of Britain and Ireland (Plate 1.2). For the purposes of this Section, we are going to imagine that, at the scales of this map, the ground surface can be treated essentially as a horizontal plane. For example, at the scale of the Ten Mile Map (1 : 625 000), a mountain 625 000 cm (i.e. 6250 m) high would be only 1 cm high on a true-scale topographic profile across the map. Ben Nevis, which is approximately 1343 m (4406 feet) high, would be only about (1343 × 100) / 625 000 = 0.21 cm high, and less than one-tenth of that at the scale of the colour plate map (1 : 7 000 000). Remember that the direction in which outcrops of strata are elongated will be parallel to the strike of the beds, and the direction of dip is at right angles to this strike direction.

2.6.1 DIPPING STRATA

Look on Plate 1.2 at the rocks of the Jurassic; they form a single large (olive-green) band across England. All along the south-east of the Jurassic outcrop, the adjacent rocks are Cretaceous (pale-green) and to the north-west are Triassic (cream).

❏ From this, can you tell what is the general direction of dip for the Jurassic rocks of England as a whole?

■ Since younger beds occur from the north-west to the south-east across the Jurassic outcrop, they must be dipping to the south-east. We can say that the *regional dip* of the Jurassic rocks of England is to the south-east. For exactly the same reason, the regional dip of the Cretaceous rocks north of London is in a similar direction.

2.6.2 Folded strata

Again on Plate 1.2 look at the outcrop pattern of the Cretaceous and Tertiary rocks in south-east England. You have already seen that there is a trough or synclinal fold in the Hampshire Basin (Section 2.3.4) where younger Tertiary strata are virtually completely surrounded by older Cretaceous strata.

❏ What kind of structure lies below London?

■ You have seen that the Jurassic strata north of London are dipping to the south-east and are overlain by younger Cretaceous beds. London lies on even younger Tertiary beds, and south of London the Cretaceous reappears. Therefore London also must lie in a synclinal fold. This is known as the London Basin.

❏ Can you tell from the map what kind of structure lies due south of London?

■ No. All the beds here are Cretaceous (pale-green) in age.

 Now look at the Ten Mile Map south of London: the Cretaceous strata are subdivided into five separate units (102–106).

Question 2.13   Consider a journey from London (TQ (51) 37) via East Grinstead (TQ (51) 3937) to Brighton (TQ (51) 3205) over the Weald area of Sussex and Kent. Look at the sequence of Cretaceous strata over which you would travel.

(a) What is the general strike direction of the beds between London and East Grinstead?

(b) Do you cross on to older or younger beds going from London to East Grinstead? In which direction do they dip?

(c) Do you cross on to older or younger beds going from East Grinstead to Brighton? Do they dip in the same direction as the beds to the north of East Grinstead?

(d) On Figure 2.23, which represents the land surface profile from London to Brighton, roughly sketch in the direction of dip of the beds. Since the scale here is 1 : 625 000, do not try to show the strata accurately. Can you see what structure underlies the Weald?

Figure 2.23   Section line from London to Brighton. (For use with Question 2.13.)

You should now have a good idea of the main structural features of south-east England. The Mesozoic strata south-east of a line from Dorset to East Anglia form three major folds: the synclines of the Hampshire and London Basins are separated by the anticline of the Weald. Within this broad structure there are also many smaller folds down to the scale of a few tens of metres as, for example, shown in Plate 2.3. As we go north-westwards on to rocks older than the Jurassic, the pattern of folding becomes more complex.

2.6.3 UNCONFORMITIES

Unconformities, particularly ones which extend over a large area, can also be recognized from map evidence. Remember from Section 2.4.2 that unconformities represent stratigraphic gaps, so the stratigraphic sequence is incomplete. Unconformities often represent major events in the geological history of an area, and major unconformities often correlate with plate tectonic events. As at Ingleton, an unconformity is often a more or less horizontal plane, or dips at a shallow angle, cutting across more steeply dipping strata below.

Find Kendal (SD (34) 5293) on the Ten Mile Map (S) and examine the boundary between the Silurian (74) and Carboniferous (80) strata.

❑ How would you describe the boundary? Is it straight and linear or curved and indented?

■ It is curved and indented. This gives us a clue as to how to distinguish between unconformities and faults on a map.

In the Kendal area, the whole of the Devonian Period is missing, representing a huge gap in the geological record. But the junction *could* be a fault (as we saw at Ingleton, which is nearby). However, whereas the fault line at Ingleton was almost straight, the boundary near Kendal is indented: so this boundary marks a horizontal or gently dipping unconformity like that in Chapel-le-Dale.

2.6.4 FAULTS

Only the very largest faults, of the order of several tens of kilometres long, are shown on the Ten Mile Map.

❑ On the Ten Mile Map (S), look at the part of the long fault that runs north-east to south-west through SJ (33) 50. (Confine your attention to the south-western part of this 10 km square.) To what period do the rocks (a) to the north-west and (b) to the south-east of the fault belong?

■ Rocks of the Carboniferous Period (84) lie to the north-west, and rocks of the Ordovician Period (70) lie to the south-east.

The difference in age of these two sets of rocks is more than 100 million years; there has been a major movement on this fault to bring the much older rocks in contact with the younger ones. This fault is the Church Stretton Fault, named after the town at SO (32) 4593.

There are several major faults on the Ten Mile Map (N), some of them over 100 km long. The Southern Uplands Fault runs from the entrance to Loch Ryan (NX (25) 07) to Dunbar (NT (36) 6879). The Highland Boundary Fault runs from Helensburgh (NS (26) 2983) to Stonehaven (NO (37) 8786) and marks the south-eastern edge of the Scottish Highlands. You will see that it separates mountainous ground consisting of metamorphic rocks of the Dalradian Complex (13–25) from younger sedimentary rocks of Devonian age (75). The Great Glen Fault runs south-west from Inverness (NH (28) 64) to the Isle of Mull (NM (17)), cutting rocks of the Moine Complex (8–12).

There are also examples of major thrusts on this map, the largest of which is the thrust running up the coastal strip of north-west Scotland, from Point of Sleat on Skye (NG (18) 5600) to Whiten Head (NC (29) 5069), on the north coast. This is called the Moine Thrust. The rocks of the Moine Complex have been thrust north-westwards over the *younger* Cambrian and Ordovician sediments (62, 63 and 67). You will learn more about the formation of these major structures in Block 3.

2.6.5 DYKES

We now return to the regular pattern of dykes on the Ten Mile Map (N) near NS (26) 09 that we mentioned in Section 1.5. Such a pattern is called a **dyke swarm.** Some of the dykes extend beyond the Dalradian, and can be traced south-eastwards across southern Scotland and into England. One of these dykes (33) runs through Armathwaite (NY (35) 5047) in northern Cumbria.

❑   What can you deduce about the age of the Armathwaite dyke?

■   The dyke cuts boundaries between Permian and Triassic rocks, both to the south-east and to the north-west of Armathwaite. Therefore, it is at least younger than these rocks, and its *maximum* age cannot be more than about 200 million years.

But perhaps the dyke swarm was much younger than this. If you now look very closely at southern Skye (NG (18) 51) you will see outcrops of Middle Jurassic rocks (94–96) and of a pale-green formation which, although unlabelled, is actually Upper Chalk (106). The boundaries between the Chalk and the Tertiary basalt (57) here are cut by the dyke swarm, so the dykes are *even younger* than the Cretaceous. The swarm is in fact of Tertiary age, though later than the Tertiary basalts which the dykes also cut. This last example serves to illustrate the kind of detail which may be extracted from the Ten Mile Map.

2.6.6 THE GEOLOGICAL STRUCTURE OF BRITAIN

It is often helpful to bring together the most important geological features of an area by drawing a simple map that shows the boundaries between rocks belonging to the major eras, the main igneous and metamorphic rocks, the major faults and thrusts and major unconformities. Such a map is called a structural geological map.

Figure 2.24   Outline map of the British Isles showing major structural boundaries. (For use with Question 2.14.)

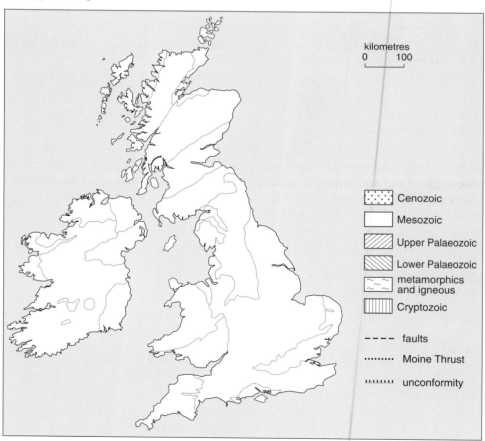

Question 2.14   Using the Geological Map of Britain and Ireland (Plate 1.2):

(a) Indicate on Figure 2.24 the areas occupied by the various groups of rocks shown in the key.

(b) Mark the major faults and thrusts as shown in the key (you will need to use the Ten Mile Map).

(c) Mark in and label the unconformity at the base of the Mesozoic rocks.

You have just completed a crude structural map of the British Isles which is, in effect, a simplified version of the colour plate map. As you work through the rest of this Block, you may find it helpful to mark areas covered by each of the BGS maps on Figure 2.24. Both Moreton-in-Marsh (Section 3) and Cheddar (Section 4) lie within the broad belt of Mesozoic strata, but as you should have already realized from the Ten Mile Map, there are many minor complications to the geology which become obvious only on a larger-scale map.

2.7 SUMMARY OF SECTION 2

- The shape of the land surface is commonly controlled by the underlying geology.
- The geologist can use information from landforms together with isolated rock exposures to interpret the underlying geology.
- Deformation of rocks results in folding and/or faulting of the rocks.
- Geological structures can be interpreted from the pattern of rock outcrops that appears on a geological map.

2.8 OBJECTIVES FOR SECTION 2

Now you have completed this Section, you should be able to:

2.1   Draw simple topographic profiles across Ordnance Survey maps and sketch simple geological cross-sections across areas of the Ten Mile Map.

2.2   Using evidence from geological maps and field examples, recognize features such as exposures of solid rock, breaks of slope and spring lines, and infer the presence of geological structures such as boundaries between different rock types and dip and scarp slopes.

2.3   Explain the relationship between the dip and strike of beds and between the angle of dip and width of outcrop on simple geological maps, and deduce the regional dip and strike of an area from the outcrop pattern and the stratigraphic column.

2.4   Recognize major fold structures and correctly identify the type of fold as anticline or syncline from outcrop patterns on the Ten Mile Map.

2.5   Recognize major fault structures and simple igneous intrusions such as dykes and plutons from their outcrop pattern on the Ten Mile Map.

Now try the following questions to test your understanding of Section 2.

Question 2.15   On the Ten Mile Map (S), look at the Carboniferous rocks of South Wales (beds 80–83), in particular along the Eastings grid line 320 north from Cardiff (ST (31) 2075) to the River Usk (SO(32) 2018).

(a) Which way are the strata (beds 80–83) dipping between north of Cardiff and Caerphilly (ST (31) 1687)?

(b) Which way are the strata dipping south of the River Usk?

(c) What structure occurs between Cardiff and the River Usk?

(d) What is the general strike direction of the country between Ammanford (SN (22) 6312) and Brynmawr (SO (32) 1912)?

(e) What is the strike direction between Margam (SS (21) 8086) and Caerphilly?

Now look at the outcrop pattern between Risca (ST (31) 2491) and Blaenavon (SO (32) 2508).

(f) What is the general strike direction in this area?

(g) What is the dip direction in this area?

 Question 2.16   You have met some of the rock types depicted in the South Wales area before: the Carboniferous Limestone Series and Millstone Grit Series. The Old Red Sandstone and the Pennant Measures are both well-cemented sandstones that were deposited under water.

(a) What topographic features would you expect to see when travelling south from Brecon (SO (32) 0528) to Cardiff?

(b) Figure 2.25 is a cross-sectional sketch showing topographic features and indicating some of the strata outcrop. Complete this cross-section by drawing in the boundaries between 83 and 80–82 and between 80–82 and 75.

Figure 2.25   For use with Question 2.16(b).

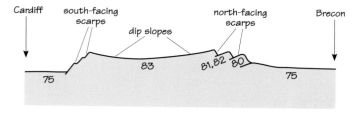

 Question 2.17   Describe, and suggest two possible reasons for, the differing width of outcrop on the north and south limbs of the South Wales coalfield.

 Question 2.18   Look at the area between Alton (SU (41) 7139) and Haslemere (SU (41) 9032). What are the strike and dip of the strata in this area?

 Question 2.19   Examine the north–south fault lying between Birmingham and Coventry that passes near Meriden (SP (42) 2582).

(a) What strata are shown to the west of the fault?

(b) What strata are shown to the east of the fault?

(c) If this is a normal fault, on which side of the fault have the beds moved down?

3 THE MORETON-IN-MARSH SHEET

Throughout this Section, you will be making frequent use of the Moreton-in-Marsh map Sheet. You will also need your Ten Mile Map (S) and Multimedia access at various points.

3.1 INTRODUCTION

As outlined in Section 1.4.3, a geological map should be thought of as a representation of the pattern of rocks cropping out at the Earth's surface. However, most geological maps contain much more information than simply patterns of surface outcrop. For example, on the one hand they can be used to build up a detailed three-dimensional picture illustrating the relationships of the different rock units in the uppermost layers of the Earth's crust, whilst on the other they provide interpretative tools in the form of special symbols and colours which can help the user to determine the geological history of the region. The approach of extracting information illustrated in the following Sections provides a general methodology which you can use to help you understand more complicated examples of geological map sheets.

The objective of this Section is to demonstrate how you can use the range of geological information available on a map sheet to understand the relationship between geology and topography, and then to build up a geological history outlining the sequence and pattern of formation or deposition, faulting, and erosion. You will first be shown how to identify the characteristic patterns produced by horizontal strata on geological maps, and then how to work out the relative vertical displacement of strata across faults. Finally, you will be asked to draw geological cross-sections and compile a geological history of the area.

3.2 THE MAP

The Moreton-in-Marsh Sheet offers an excellent introduction to 1 : 50 000 maps typical of those used by geologists. This map shows an area which is largely unaffected by folding and therefore the near flat-lying nature of the rock strata will allow you to examine in detail the relationships between the patterns of outcrop and local topography. The differing resistances to weathering and erosion of the various rock strata in the Moreton-in-Marsh area have exerted fundamental controls upon the development of the present-day topography.

First, establish the position of the Moreton-in-Marsh Sheet on the smaller-scale Ten Mile Map (S). It lies in the 100 km grid squares SO (32) and SP (42) and forms part of the region at the edge of the Cotswold Hills lying north-east of Cheltenham. You will also find the outline of the Sheet (No. 217) marked in red on the Ten Mile Map. From the Ten Mile Map, you can also see that the Moreton-in-Marsh area forms a segment of the wide Jurassic outcrop (brown and yellow colours, 91–101) which generally trends north-east to south-west across England. These Jurassic strata belong to part of the Mesozoic succession of sedimentary rocks that underlie the south-eastern half of Britain and which generally dip towards the south-east. However, this dip is typically small (about 1–3°), so that on relatively large-scale maps, such as the Moreton-in-Marsh Sheet, the effect of low dip angles on the strata is difficult to see and hence they appear almost horizontal. Outcrops of Jurassic rocks can be followed on the Ten Mile Map (S) from Lyme Regis on the south coast, throughout Somerset, Worcestershire, Oxfordshire, and through more northerly counties including

Northamptonshire, Lincolnshire, and finally into North Yorkshire where they form coastal outcrops near Scarborough (TA (54) 0487) and underlie the moorland country inland. Topographically, the more resistant Middle Jurassic lithologies form a belt of rolling hills, whilst the underlying softer Triassic and Lower Jurassic rocks tend to form valley floors and low-lying areas.

Now that you have established the Moreton-in-Marsh area in its regional geological context, we can consider the information given on the sheet itself.

3.2.1   General layout of the map sheet

Large-scale BGS geological maps such as the Moreton-in-Marsh Sheet typically comprise three basic components. These are (1) the map itself, (2) areas in the side margins of the map illustrating a geological column and the key to symbols used, and (3) a cross-section through the geology of the area which is usually printed in the lower margins of the sheet (on more modern maps, this is often accompanied by brief geological notes). When working with any geological sheet, you should first look at the material in those margins since they provide the information which will tell you how to use and interpret the map. A useful approach is to examine in turn each piece of information provided on the sheet. Therefore, the aim of Sections 3.2.2–3.2.8 is to show the importance of examining and understanding the marginal information (i.e. the key and cross-section) on a sheet before starting to look closely at the map itself. Section 3.7 provides a useful checklist of the types of geographical and geological information that should be examined when you first look at a geological map. The remaining items in the margin of the map include an index to the Six-Inch geological sheets of this area, and another to the adjacent 1 : 50 000 or One-Inch sheets.

3.2.2   Index and key to formational symbols and colours

The title and scale appear in the top right-hand blue box together with background data outlining how and when the geological and topographic information were compiled. For the Moreton-in-Marsh area, there are two important things to note here: (1) the topographic base map is a metric one, so contours are in metres, not feet; and (2) the map is a Solid and Drift edition which means that superficial surface deposits such as Boulder Clay, landslipped materials, river gravels and sands (alluvium) which conceal the underlying solid strata (these are collectively termed 'drift'), are also shown. The Moreton-in-Marsh Sheet has the modern layout style of British geological maps because it was reprinted in 1981 with revised marginal information.

The Moreton-in-Marsh Sheet is a combined Solid *and* Drift edition because the small areas of drift do not appreciably obscure the solid geology below. However, in some parts of Britain, especially parts of Scotland, where the drift is thick and very extensive, separate Solid and Drift sheets are printed.

The blue box below the title box on the right of the map contains the 'Index and explanation of formational symbols and colours' for the drift materials and the solid geology, and also the key to the map symbols which are necessary and relevant for the construction and interpretation of this particular sheet. This information tells us what kinds of surface deposits (drift), rock types (solid) and their ages, we are to expect within the area of the map. As a consequence, the information contained in this index is unique to each different map sheet.

The index for the drift deposits shows the different kinds of superficial deposits present. These include an extremely varied series of unconsolidated beds of Quaternary age such as sands, gravels, clays, river alluvium and Boulder Clay which were left after melting of ice-sheets. On the Moreton-in-Marsh Sheet, it also includes landslipped material.

Generally, in this Course we will be concerned primarily with the solid geology. This is shown diagrammatically as a stratigraphic column and indicates the pre-Quaternary rocks present at or near the surface, together with their relative ages, although in many places these may be covered by drift, soil and vegetation. The colours used for the different rock types follow a convention and so tend to be similar for rocks with the same age and lithology on all BGS maps. Using this system, the Lower–Middle Jurassic rocks are given mostly yellows, oranges and browns and, therefore, their presence can be easily recognized on both the Moreton-in-Marsh Sheet and on the Ten Mile Map (S). Clearly, since more detail is available on the Moreton-in-Marsh Sheet and more rock units can therefore be recognized, a wider range of colours and colour shades is needed to present the detailed stratigraphic information. Different types of drift deposit are also given their own colours and symbols.

Immediately below the solid geology column, there is a key to the symbols that describe the direction and amount of dip of the strata. Horizontal beds are indicated by a cross. Where rock strata have a measurable dip, the direction and amount of dip are shown by an arrow and a value in degrees (on older BGS or IGS maps this may be signified by a bar with a tick mark showing dip direction). On this particular map, a larger arrow (without dip amount) is also used to indicate general dip in an area where individual orientations and values become very variable but where an overall dip direction can still be recognized.

❑   Where on this map has the larger arrow symbol been used?

■   The rocks in the north-west corner of the map, the outcrop of Inferior Oolite, (InO), is marked as having a general south-westerly dip. Individual values here may be variable due to the effects of local landslippage.

The key also shows the different symbols used to represent geological boundaries. A solid line is used where a geological boundary can be determined with certainty; for instance, where stratigraphic boundaries are exposed in quarries, in cliff faces, or otherwise obviously expressed in the topography. Dotted lines are used for geological boundaries that cannot be accurately positioned on the ground due to poor exposure. Faults are typically marked as a heavy broken line, though they are usually shown as a solid heavy line where they become exposed. Most importantly, a small tick mark indicates the side of relative downwards movement (e.g. the tick mark at SP 016232 indicates relative downwards movement on the southern side of this fault).

3.2.3 SOLID GEOLOGY

Each of the coloured boxes in the 'Index and explanation of formational symbols and colours' represents a conveniently mappable stratigraphic unit with the youngest at the top and progressively older ones below. On the Moreton-in-Marsh Sheet, the different rocks are denoted not only by different colours but also by letters, usually the initials of the rock name, for example, LLi = Lower Lias. Some of the names given to the stratigraphic units are the same as those used on the Ten Mile Map, for example, Great Oolite (95), Inferior Oolite (94) and the Upper, Middle and Lower Lias (93–91). Brief descriptions of the different rock types are included in this solid geology column. It is worth noting that in areas of Jurassic rocks beyond that shown on this map, the number and character of the rock units may vary due to differing depositional conditions.

3.2.4 THE DRIFT DEPOSITS

In the solid geology column, the rocks are shown in order of their age (youngest at the top). However, because drift deposits are often discontinuous and their relative ages cannot always be precisely determined, the drift column might not be in strict stratigraphic order. The last paragraph of 'Notes on the geology' at the foot of the Moreton-in-Marsh Sheet reveals that determining relative ages for the different drift deposits has been a problem in this area.

Boulder Clay, a deposit consisting of pebbles and gravel-sized particles set in clay, is evidence that an ice-sheet extended as far south as Moreton-in-Marsh during the last ice age (which ended about 10 000 years ago). It is left behind when glaciers or ice-sheets melt. The various sands and gravels (green, brown and lilac on the map) are glacial materials redistributed by rivers fed by the melting ice-sheet as it retreated northwards. Modern alluvium and river gravels (cream and light brown) are found along the valleys of any sizeable stream or river, especially where the gradient is low, for example, the Evenlode and the Isbourne (e.g. SP 2227 and SP 0337).

3.2.5 THE GENERALIZED VERTICAL SECTION

On first glance, you may think that the generalized vertical section in the bottom right of the sheet simply repeats information given in the solid index. However, the solid column is only designed to tell you which rock types occur at the surface, but it does *not* give other crucial geological information such as the relative thicknesses of the beds, how individual beds may vary in thickness, whether there are gaps in the sedimentary succession, or whether there is information available about the subsurface rocks lying below those exposed within the map area. A generalized vertical section is given to help answer these questions.

In the generalized section, the thickness of rock units is drawn to a scale of 1 : 5 000 so that you can see at a glance their comparative thicknesses. Look at the *range* of thicknesses given for each unit. For instance, the Upper Lias (ULi) ranges from 25–120 m in thickness, and the Chipping Norton Limestone (CNL) from 0–9 m, and so average or typical values have been used to construct the vertical section. You will also see that where the thickness ranges down to zero, the unit is usually shown as a wedge in the stratigraphic column.

Detailed information regarding subsurface rocks has been determined from core material retrieved from the Lower Lemington Borehole, and has been supplemented by other borehole information as described in the 'Notes on the Geology'. This gives further information about the geology beneath the rocks exposed in the Moreton-in-Marsh area and which can be added to that available from the solid index. Borehole data show that the Lower Lias (LLi) succession, representing the oldest rocks exposed at the surface (typically 170–300 m thick in this area), are underlain by progressively older successions from the Triassic, Carboniferous and Silurian Periods.

Unconformities (Section 2.4.2), representing major gaps in the sedimentary succession, are shown as breaks with wavy lines in the generalized section. The lowermost shows that a major gap exists between late Silurian (Llandovery age) rocks and the overlying Upper Carboniferous (Westphalian age) rocks. The uppermost shows a gap between the Upper Coal Measures of the Carboniferous and the Triassic Bromsgrove Sandstone.

❏   Using Figure 1.2, which part of the geological column is absent at this uppermost unconformity? Make an estimate of the amount of time unrepresented in this part of the succession.

■   There are no Permian rocks detected in the borehole, so there must be at least 42 Ma unrepresented in the succession in order to account for the missing Permian. There are two possible reasons for this. It is possible that Permian rocks were originally lain down, but then eroded away prior to the deposition of the Triassic strata; alternatively, it could be that Permian rocks were never deposited. In fact, inspection of the generalized vertical section reveals that the gap may be of even longer duration because the succession also lacks the oldest Triassic and youngest Carboniferous.

3.2.6 THE ROCKS OF THE MORETON-IN-MARSH AREA

A general overview reveals that the map consists of a roughly north–south-trending central region in which rocks of predominantly Middle Lias (MLi) and younger age occur, flanked on either side by areas of Lower Lias (LLi) upon which most of the drift deposits are located. Inspection of the topographic information (i.e. contour lines) reveals that the Lower Lias underlies the relatively flat low-lying valley areas, and that the central area comprises a NNE–SSW-trending ridge of hills which are formed of Middle Lias at the base and then overlain by progressively younger strata capped by the youngest rocks such as the Chipping Norton Limestone (CNL) and Great Oolite Limestone (GOL). A prominent area of landslippage may also be identified in the north-west corner.

The Jurassic strata of this area were laid down beneath the sea between about 200 to 165 Ma ago during the Lower and Middle parts of the Jurassic Period, which was part of the Mesozoic Era (Figure 1.2). The Lower Jurassic is divided into Lower, Middle, and Upper Lias, the name 'Lias' being an old quarrying term probably derived from a French word 'liois' meaning a type of limestone. In this area, the Lower and Upper Lias are described as mainly alternating beds of clays and thin impure limestones with some silts and sands (e.g. see the index describing the solid geology). The Middle Lias contains some limestone, much of which is very rich in iron, for example the Marlstone Rock Bed (SO 9938), which in the past provided an important source of iron ore for Britain's steel industry.

The stratigraphic units above the Middle Lias are commonly quite thick in the Cotswolds, but can be restricted in lateral extent so they tend to have been given local names. Many of these names are self-explanatory such as 'Cotteswold Sands' and 'Chipping Norton Limestone'. The Inferior and Great Oolite Limestone units predominantly contain beds of oolitic limestones, such as RS 22 in your Home Kit, which comes from a quarry in the Inferior Oolite of this area (you will examine this specimen in Block 2). Caution should be exercised when interpreting local rock names since these may refer to a property or usage of the rock, rather than its geological attributes. For instance, the name Stonesfield Slate is overprinted on some of the Great Oolite Limestone (GOL) areas in the central southern part of the map, but this is not a slate in the metamorphic sense (metamorphism is a term which you will discover more about in Block 2): in fact, it is a variety of oolitic limestone. Here the word 'slate' is taken from quarrying terminology, and simply refers to the fact the rock can be split into thin sheets suitable for roofing material. At the very top of the stratigraphic column the 'Forest Marble' is another geological misnomer since, geologically speaking, a true marble is a metamorphic rock. The Forest Marble is not metamorphic; it is a coarse-grained limestone first described from localities in Wychwood Forest (near Oxford), but is referred to as marble by quarry workers because it can be polished as an ornamental stone. It is the youngest rock exposed in this area and there are only three small outcrops of it, the largest of which is at SP 143279.

3.2.7 GEOLOGICAL CROSS-SECTION OF MORETON-IN-MARSH

The **geological cross-section** at the foot of the map sheet is a slice through the succession of rock strata (i.e. the stratigraphy) and associated topography, and is designed to show the general structure of the rocks in this area. It is an interpretation of the subsurface geology deduced from the distribution of rock units mapped at the surface. Sometimes, as in this area, evidence for the subsurface geology is augmented from borehole data. Often the cross-sections on geological maps are not drawn along a straight line, but instead have a dog's-leg path chosen to pass through areas of interest to show as much as possible of the key geological relationships. On the Moreton-in-Marsh Sheet, for instance, the section runs roughly east–west but changes direction at Cutsdean Hill. The grid references for the ends of the cross-section line are given above the section itself. The cross-section and the map together provide two of the main pieces of information which help geologists visualize the three-dimensional relationships of the rocks within a particular area.

The cross-section shows that the strata in this area are tilted only very slightly from their original horizontal orientation. The top of the cross-section (the topographic profile) shows the topographic irregularities of the land surface and demonstrates how hills and valleys interrupt the continuity of the rock layers. This is particularly obvious towards the western (left-hand) end of the section where the River Isbourne has eroded a wide valley in the strata down to the Lower Lias (LLi) leaving higher ground on either side composed of younger rock units.

On the eastern (right-hand) end of the cross-section, information from the Lemington borehole has been projected onto the line of section. This has provided information about rocks unexposed in the Moreton-in-Marsh area but which are presumed to underlie the Jurassic stratigraphy; for instance, the Mercia Mudstone Group (MMG). If you follow the Mercia Mudstone Group westward, you will see it becomes much thicker indicating that this western area was a sedimentary basin which received much sediment during Triassic times. Below these mudstones, even greater thicknesses of Permian and Triassic sandstones are thought to exist, but because further details are unavailable the boundaries between these units are uncertain and so are marked with dotted lines.

To the east of Cutsdean Hill, the section crosses a fault. This is shown as an inclined black line which displaces the Jurassic stratigraphy downwards to the right (east). Again, a lack of evidence of exactly how this fault may affect the Mercia Mudstones Group beneath means that it is not drawn to the bottom of the cross-section. If you now examine how the fault affects the Jurassic succession, it should be possible to estimate the vertical displacement of this fault by measuring the vertical distance between the same horizon recognizable on either side of it (e.g. top of the Csd unit). For instance, when measured directly on the section, the vertical distance between the top of the Cotteswold sands (Csd) lying east and west of the fault is 2 mm. Using the scale provided at the left-hand end of the cross-section to convert this measurement reveals that the vertical movement of the fault is about 50 m.

3.2.8 VERTICAL EXAGGERATION AND RELATIONSHIP OF MAP SCALES

With all geological map sheets, it is crucial to determine the scales used for constructing the cross-section. In many cases, the vertical and horizontal scales will not be the same. The vertical scale is often made larger, thus the section has been subjected to a **vertical exaggeration**. The reason for exaggerating the vertical scales on cross-sections is simply to provide more space in order to show all of the stratigraphic units clearly.

❑   What is the relationship between the vertical and horizontal scales on the Moreton-in-Marsh Sheet cross-section?

■   The horizontal scale is 1 : 50 000 (i.e. the map scale), and the vertical scale is *twice* this (i.e. 1 : 25 000).

In other words, the vertical exaggeration is × 2. Vertical exaggeration is the amount by which the scale of the vertical axis has been increased relative to that of the horizontal.

If the vertical scale were the same as that of the horizontal scale on the map, then the cross-section would be half as high, making it very difficult to represent

(d)

(c)

(b)

(a)

Plate 1.1   From coral seas to deserts: a reconstruction of NE England, looking westwards toward the Lake District at different geological times from the Lower Carboniferous (at the foot of the page) to the Upper Permian (at the top). The progressive northwards drift of Britain has resulted in a wide range of environments in which many different rocks were formed and later preserved.

The Lake District is composed of Ordovician and Silurian rocks with much of the high ground formed by Ordovician volcanic rocks. During the Lower Carboniferous (a), the region lay close to the equator and the lower ground to the east was covered by a shallow sea in which corals and other marine animals flourished. As this sea receded, it was replaced in turn by coal swamps in the Upper Carboniferous (b) and then deserts in the Lower Permian (c) as the British Isles region moved north from the equator and approached the present latitude of the Sahara. By the Upper Permian (d), a shallow sea had encroached once more.

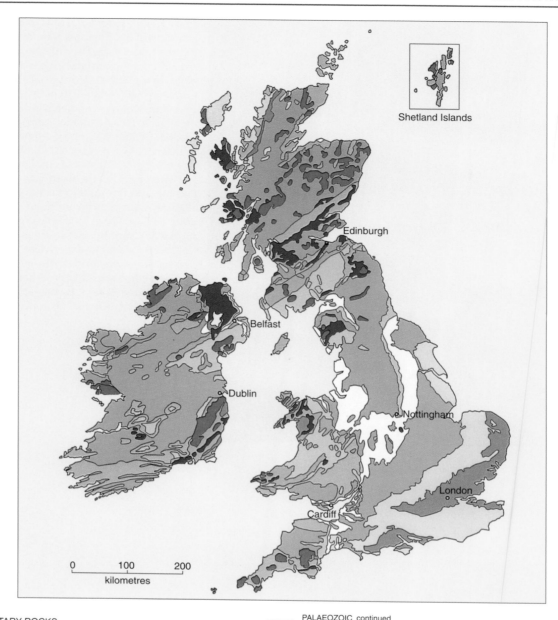

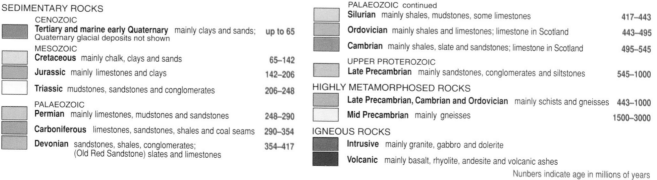

SEDIMENTARY ROCKS

CENOZOIC

Tertiary and marine early Quaternary mainly clays and sands; up to 65
Quaternary glacial deposits not shown

MESOZOIC

Cretaceous mainly chalk, clays and sands | 65–142

Jurassic mainly limestones and clays | 142–206

Triassic mudstones, sandstones and conglomerates | 206–248

PALAEOZOIC

Permian mainly limestones, mudstones and sandstones | 248–290

Carboniferous limestones, sandstones, shales and coal seams | 290–354

Devonian sandstones, shales, conglomerates; | 354–417
(Old Red Sandstone) slates and limestones

PALAEOZOIC continued

Silurian mainly shales, mudstones, some limestones | 417–443

Ordovician mainly shales and limestones; limestone in Scotland | 443–495

Cambrian mainly shales, slate and sandstones; limestone in Scotland | 495–545

UPPER PROTEROZOIC

Late Precambrian mainly sandstones, conglomerates and siltstones | 545–1000

HIGHLY METAMORPHOSED ROCKS

Late Precambrian, Cambrian and Ordovician mainly schists and gneisses | 443–1000

Mid Precambrian mainly gneisses | 1500–3000

IGNEOUS ROCKS

Intrusive mainly granite, gabbro and dolerite

Volcanic mainly basalt, rhyolite, andesite and volcanic ashes

Numbers indicate age in millions of years

Plate 1.2 Geological map of Britain and Ireland. (BD/IPR/7-14 British Geological Survey. © NERC all rights reserved.)

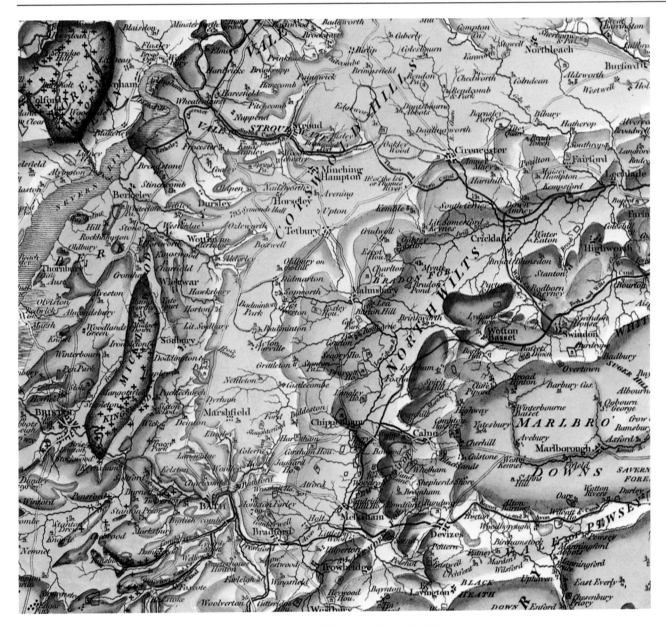

Plate 1.3   An extract from William Smith's Geological Map published in 1815.

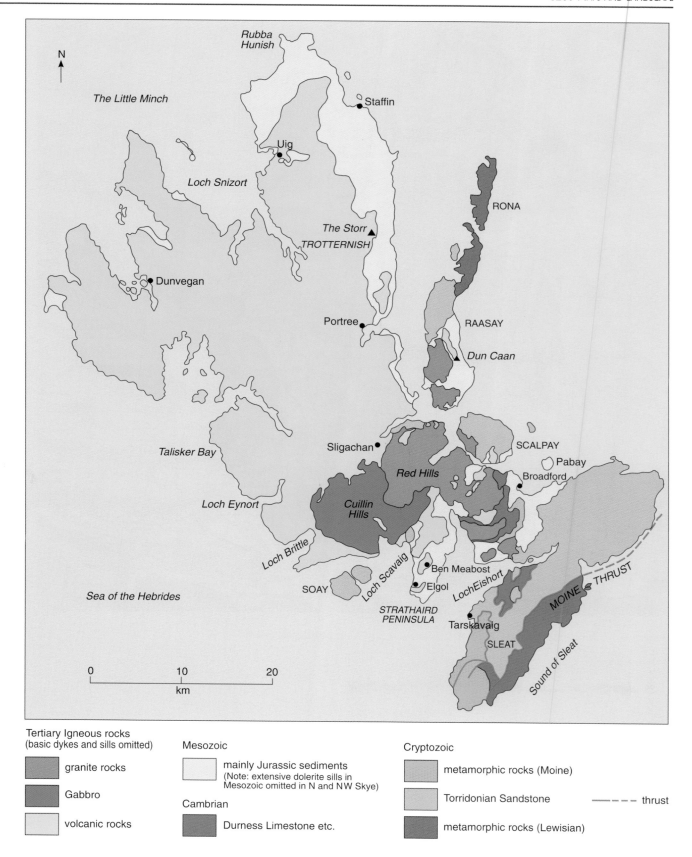

Plate 2.1   A simplified geological map of the Isle of Skye.

Plate 2.2 Panoramic view of the topography around Torrin on Skye. The granites form the prominent Red Hills to the north. The photograph is taken standing on the Jurassic sedimentary rocks around Torrin. To the west lie the Cuillin mountains, also composed of igneous rocks (gabbros) that are resistant to erosion.

Plate 2.3   Folds in the Purbeck Beds at Stair Hole, Lulworth Cove, Dorset. The axial planes are inclined, dipping to the right so that strata on opposite sides of the axial planes dip at different angles. (*Nick Watson, OUPC/BBC*)

Plate 2.4   Syncline in the Chisos Mountains of Texas, USA. Folding of rocks can occur over a wide range of scales from a few centimetres to hundreds of kilometres.

Plate 2.5 The outcrop at Siccar Point, Hutton's classic unconformity. Gently dipping rocks of Upper Devonian age lie unconformably on near vertically dipping Silurian-aged rocks. Inset shows dull-red Devonian sandstones on top of grey Silurian shales. (*Dave Williams, Open University*)

thinner stratigraphic units such as the Chipping Norton Limestone and Cotteswold Sands. However, the disadvantage of using vertical exaggeration is that it produces a steepening of the dip of any inclined strata drawn on the cross-section; this causes serious problems because you cannot then measure accurately dip or thickness directly from the section. This distorting effect should always be borne in mind since it may become necessary to use large vertical exaggeration (e.g. × 3 or × 4) when you draw sections across more complicated maps. However, on larger-scale maps, such as the 1 : 25 000 series (e.g. Cheddar Sheet in your Home Kit), it does become feasible to present the data at a similar vertical and horizontal scale, thus avoiding distortion. Similar horizontal and vertical scales are required for the construction of **accurate cross-sections**.

3.3 USING THE MAP TO INVESTIGATE STRATIGRAPHY AND STRUCTURE

Remember from Section 1.4.3 that a geological map such as the Moreton-in-Marsh Sheet presents an interpretation of the surface geology as it would look if all the soil and vegetation were stripped off; in other words, where the different rock units would form *outcrops* at the surface. In practice, actual *exposures* where the rocks can be seen at the surface are very limited in this area due to soil cover, and geologists have instead had to rely on a variety of mapping techniques (i.e. changes in soil type, vegetation, or topographic features) and data sources (e.g. the Lemington borehole) to produce the final map.

3.3.1 IDENTIFYING THE EVIDENCE FOR HORIZONTAL STRATA

Although we can see from the cross-section and the presence of the cross symbol (+) denoting horizontal strata that most of the rocks in the region are flat lying, or nearly so, a better way of demonstrating this over much of the map area is by examining the relationship between outcrop patterns and topography. We shall begin by looking at the outcrop patterns of three units, the Middle and Upper Lias (MLi and ULi) and the Cotteswold Sands (Csd).

❑ First, find Broadway at SP 0937 and Winchcombe at SP 0228, and look at the outcrop pattern of these three stratigraphic units between these places. What is the relationship between the outcrop pattern and the contours along this strip? If you have difficulty finding the contours, it may be useful to use a hand lens (there is one in your Home Kit).

■ The three stratigraphic units form parallel strips which follow the bulges and indentations of the contour pattern. In other words, the stratigraphic boundaries between these units are all *more or less parallel* to the contours.

If you now look in the south-east corner of the map around Wyck Beacon (SP 2020), you will find a similar relationship between geological boundaries and topography, suggesting this pattern holds true throughout the whole map.

We saw in Section 1.4.2 that contours represent lines joining all points of equal height above sea-level; thus each contour on the map can be imagined as an intersection between the topography and an imaginary horizontal plane at a specific height above mean sea-level. Since geological boundaries on this map (i.e. the lines where the tops and bottoms of stratigraphic units intersect the ground surface) also follow these topographic contours, this tells us that the stratigraphic units must also be horizontal. This can be summarized as follows:

Where geological boundaries between strata are parallel to the topographic contours, the strata *must* be horizontal.

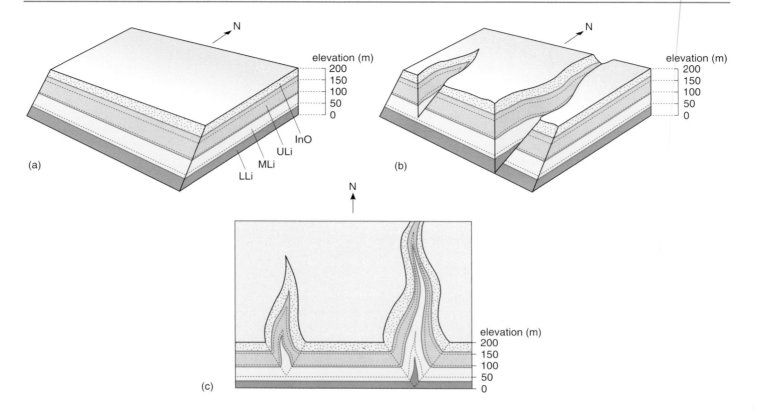

Figure 3.1 The layer cake stratigraphic model. (a) Perspective view of layer-cake model. The boundaries between layers are parallel and horizontal, and so where they intersect the front (escarpment) edge they are parallel to the contour lines which are shown as dashed lines. (b) Two valleys are cut in the model producing outcrop patterns which form a V-pattern up each valley. (c) A plan view of the model and valleys as in (b). Outcrops of the layers and contours 'V' up the valley.

Figure 3.1 illustrates this point. In Figure 3.1a, the layer-cake stratigraphy displayed at the front of the block model is comparable to the escarpment edge you observed east of Winchcombe and Broadway. Cutting into this layer-cake model (Figure 3.1b) would be analogous to a river eroding a steep valley in the succession of horizontal strata and the boundaries between the layers would also form horizontal lines on the cut surfaces making a V-pattern that points up the valley. This V-pattern is illustrated best in the plan view (Figure 3.1c) and presents the same perspective as that of a geological map.

3.3.2 OUTCROP V-PATTERNS

V-patterns in the outcrop pattern also provide another very important clue to geological structure because they tell us a great deal about the dip of strata. You already know that contours go round hills and ridges, and form Vs that point upstream (Activity 2.2), and you now *also* know that the outcrops of horizontal strata must *always* be parallel to contours (Figure 3.1c). Accordingly, a general rule follows:

> Outcrops of horizontal or gently inclined strata usually form prominent V-shaped patterns that point upstream in valleys.

Now look at the outcrop pattern of Middle Lias (MLi), Upper Lias (ULi), and Cotteswold Sands (Csd) along the hills east and south-east of the villages of Hailes and Stanway between SP 0732 and 0528. Here, two valleys have cut into the Jurassic stratigraphy producing patterns similar to that shown in Figure 3.1c. The presence of these patterns confirms that the strata here are near horizontal.

3.3.3 RELATIONSHIP BETWEEN HORIZONTAL STRATA AND TOPOGRAPHIC SLOPE

The two most important factors responsible for the outcrop patterns of horizontal (or gently inclined) strata on this or any other map are topographic slope and thickness of the strata.

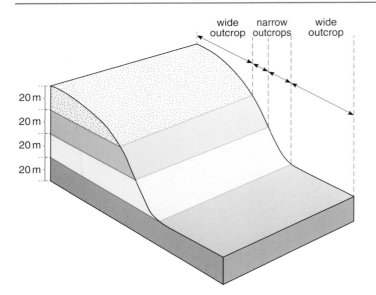

Figure 3.2   Block diagram illustrating how width of outcrop of beds of a similar thickness can vary according to topographic slope.

If you follow the outcrops of MLi and ULi around the edge of the Cotswold Hills which form the central part of the map, you will observe that their outcrop width on the map is not always constant: for instance, their width is narrow east of Prestbury (e.g. near SO 9823) and much wider south-west of Aston Magna (e.g. around SP 1834). One way of producing this effect would be if these beds were thicker in some areas. Where they became thicker, we might then expect to see a wider outcrop on the map. However, we can see from the cross-section that the thickness of these two units does not vary much across this central region. In a similar fashion, we can also see that the **width of outcrop** of Inferior Oolite (InO) on the map is much greater than that of the ULi, yet the generalized vertical section indicates that on average these two units are of a similar thickness. Moreover, if you now look at the Chipping Norton Limestone (CNL) you can see that it forms wide outcrops near the hill summits, but the generalized section shows it to be comparatively thin. Clearly, in these cases variation in bed thickness cannot explain the observed variation in the width of outcrop. However, the other important factor affecting width of outcrop on a geological map is *topographic slope* (Figure 3.2).

❏   What can you say about the spacing of topographic contours that lie within the outcrop patterns of the MLi and ULi near the localities of Prestbury and Aston Magna?

◼   The contours marked within these two units near Prestbury are close together indicating a steep slope, whilst those near Aston Magna are more spaced indicating a more gentle slope.

Accordingly, we can formulate another general rule:

For horizontal or gently dipping strata of constant thickness, narrow strip-like outcrops occur on steep slopes and wide outcrops occur on gentle slopes (Figure 3.2).

To verify this, now look at the western (left-hand) end of the cross-section. The outcrop width of Middle Lias (MLi) near the base of the scarp of Cleeve Cloud is relatively narrow (about 0.4 km measured along the line of section on the map itself). On the other side of the ridge, around Postlip Warren, the general slope is much gentler and the outcrop is well over 1 km wide. But the thickness of the Middle Lias is actually the same at both localities, as shown on the cross-section.

The observation that strata do maintain a constant thickness at different localities is important because similar situations may commonly be found on other geological maps. It is significant because where rock units do maintain a constant thickness, the top and bottom boundaries of that unit will be parallel. These parallel boundaries become particularly important when constructing accurate cross-sections because it is good practice for individual layers of rock to be shown maintaining a constant thickness, especially in those cross-sections which depict rock units that have been affected by faulting and folding.

3.3.4 THINNING STRATA AND LATERAL VARIATIONS IN LITHOLOGY

Whilst it is often very convenient to think of sedimentary strata as stacked layers of constant thickness, in reality this can only ever be true over limited areas. This is because if we consider a much wider area, it becomes more likely that conditions of deposition would have varied in such a manner that different thicknesses of a particular sedimentary unit were lain down at different localities. In such cases, the thickness of some units can vary considerably even within the area of a geological map, and where they thin and disappear altogether, the succession is said to display a **wedging out** of strata (Figure 3.3).

Fortunately, we can usually identify which units exhibit these thickness variations from information given on the map sheet. If you look at the written information given alongside the generalized vertical section (Section 3.2.5), you will see that considerable variations in bed thickness occur in the Moreton-in-Marsh area.

For instance, inspection of the map shows that the Cotteswold Sands (Csd) are missing in the north-western part of the area. Their thickness decreases to zero north of Cleeve Cloud (SO 984260) and they are not found in any of the hills immediately to the north or west. If you look at the cross-section, you can see that this unit is drawn as a wedge shape beneath Cleeve Cloud, indicating that it crops out on the scarp slope to the west, but not on the dip slope to the east.

Other examples include the two thin stratigraphic units below the Great Oolite Limestone (GOL), the Fuller's Earth (FE) and the Chipping Norton Limestone (CNL), which occur only in some areas of the Moreton-in-Marsh Sheet. The Chipping Norton Limestone (CNL) is widespread in the central part of the area, but wedges out towards the south-west. You can see the CNL wedging out at SP 138203; south and west of this locality it is absent. By contrast, the Fuller's Earth (FE) must wedge out toward the north-east from plentiful outcrops in the central southern part of the area.

Figure 3.3   Block diagram illustrating 'wedging out' of strata. Note that beds b and c are laterally equivalent and pass into each other along the east–west direction (in a similar fashion to the Chipping Norton Limestone and Fuller's Earth on the Moreton-in-Marsh Sheet). However, toward the north *both* beds wedge out and hence bed d eventually lies directly upon bed a. The vertical sections 1–4 demonstrate how the sedimentary succession varies at different localities as a result of this wedging out.

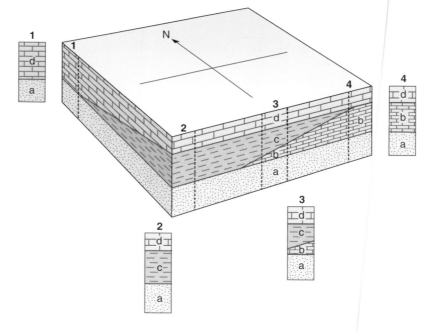

Another common characteristic of individual sedimentary units is that whilst they represent materials formed at the same time, variations in depositional conditions, and also in the proximity and type of the sediment source, often result in changes in lithology. For instance, both mudstones and sandstones are deposited by rivers, but the proportion of these varies along the course of the river so we might expect to see mudstones deposited at one locality whilst, during the same period of time, sandstones were deposited at another. Where the river enters the sea, the lithology may change again and we observe evidence of marine influence in the resulting sediment. In a geological succession, these marine-influenced sediments can be sufficiently different to those of the river that they too represent a different lithology, despite the fact that they are temporally equivalent. Often, sediments which have formed at the same time but under different types of depositional conditions grade into one another and, as a result, the composition of a single stratigraphic unit may display considerable lateral variations in its lithology. In some cases, this variation is so significant that laterally equivalent beds are given different names at different localities.

❑   How is the relationship between the Fuller's Earth (FE) and Chipping Norton Limestone (CNL) represented in the generalized vertical section?

■   Between deposition of the Inferior and Great Oolites, the Chipping Norton Limestone and Fuller's Earth were lain down. They represent two distinct lithologies which thin out laterally into one another. In other words, they were deposited at the same time under different depositional conditions. In the generalized section, they are shown as wedges indicating that their thickness decreases to zero over some parts of the area.

3.3.5   OUTLIERS AND INLIERS ON THE MORETON-IN-MARSH MAP

The concept of outliers and inliers was introduced in Section 2.4.2. Outliers occur where younger rocks are surrounded by older rocks, and inliers where older are surrounded by younger. Both can result either from unconformable relationships (as described in Section 2.4.2), *or* from faulting *or* from erosion.

On the Moreton-in-Marsh Sheet, good examples of *erosionally controlled* outliers and inliers have developed in the Jurassic strata at the edge of the Cotswold Hills. For instance, at grid reference SP 0231, you will see a small hill formed of Middle Lias (MLi). A cross-section through this hill (Figure 3.4a) shows that it

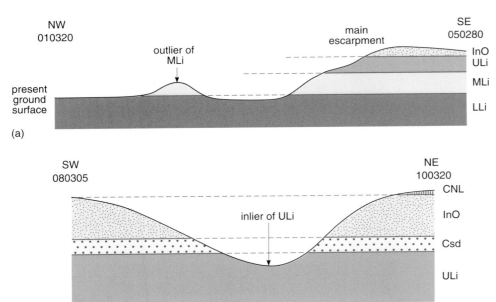

Figure 3.4   (a) Diagrammatic cross-section to illustrate the development of an outlier, such as that at SP 0231. Vertical exaggeration is about × 4. (b) Diagrammatic cross-section (SP 080305–100320) to illustrate an inlier such as that developed in the valley near Taddington (SP 088312). Vertical exaggeration is about × 10. Strata are labelled as on the Moreton-in-Marsh map.

was once part of a more extensive layer of Middle Lias sediments, but erosion has separated it from the main outcrop of the Cotswold scarp so that it now forms an isolated outcrop, or outlier, of Middle Lias. There are several other outliers in the area, e.g. Dumbleton Hill (SP 0035), Alderton Hill (SP 0034) and Woolstone Hill (SO 9731).

Now look at the long irregular strip of Cotteswold Sands (Csd) which crops out along the stream valley between SP 093330 and SP 100260. If you carefully examine the geology just east of Taddington (SP 088312), you will see that here the stream has eroded down through the Cotteswold Sands, thus making a 'window' which exposes older Upper Lias clays (ULi) in the bottom of the valley. An outcrop pattern such as this, where older rocks are entirely surrounded by younger, is a good example of an inlier (Figure 3.4b).

These examples demonstrate that the presence of an unconformity is *not* always necessary for the development of outlier and inlier outcrop patterns. Where these outcrop patterns are erosionally controlled, they may be summarized as follows:

> An *outlier* is an outcrop of rocks separated from the main outcrop by the process of erosion and surrounded entirely by *older* strata. An *inlier* is an outcrop of rock that has been exposed by erosion of younger strata in such a way that it is entirely surrounded by the *younger* strata remaining.

Now look at the inlier of Csd in grid square SP 1123. It shows a similar relationship to that above except that it is bounded to the north-west by a fault. This is, therefore, a fault-bounded inlier.

❑   How might you describe the pattern of outcrop of Fuller's Earth (FE) and Great Oolite (GOL) seen around White Hall (SP 0122)?

■   The Fuller's Earth (FE) and Great Oolite (GOL) exposed in this area are surrounded by *older* Inferior Oolite (InO) strata so they must be an outlier. In addition, to the north and south the outcrop of these younger rocks is bounded by two WNW–ESE-trending faults making this particular pattern of outcrop a *fault-bounded outlier*.

A word of caution: It is tempting to call any 'island' of younger rocks an outlier, but it would be wrong to describe all the isolated areas of drift on the Moreton-in-Marsh map as outliers. This is because many forms of drift deposits were originally only patchily distributed, and so probably never formed a continuous sheet over the whole map area. Therefore, the terms outlier and inlier are not usually used to describe superficial deposit outcrop patterns.

3.4   FAULTS ON THE MORETON-IN-MARSH SHEET

As outlined in Section 2.4.3, faults are fractures that displace rocks relative to one another. Only fairly large faults such as those more than a kilometre or so long and with displacements measurable in metres or more can be accurately shown on a map at this scale (1 : 50 000). Faults are represented on geological maps at this scale as heavy black dashed lines.

❑   Is the orientation of faults on the Moreton-in-Marsh map random, or is there a pattern?

■   Broadly speaking, there are two sets of faults present. The most obvious set is trending WNW to ESE, while a small number of faults trend roughly north to south. In several places, faults act as outcrop boundaries, especially for the FE, GOL and CNL units at the top of the stratigraphic column.

The faults on the Moreton-in-Marsh Sheet are **normal faults** (Section 2.4.3) and have resulted in one block of strata moving downward relative to another in the direction of dip of the fault (Figure 3.5a). By convention, we refer to movement

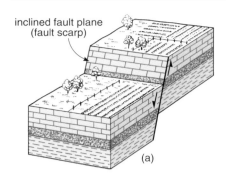
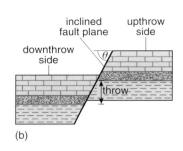
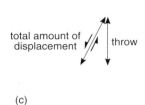

(a)   (b)   (c)

along normal faults by identifying the **downthrow** side, and when we measure the *vertical* displacement we refer to it as the amount of downthrow, or simply the throw (Figure 3.5b). Only if the fault plane is vertical (i.e. dips at 90°) can the throw of the fault be the same as the amount of displacement along the fault plane. Most normal faults dip steeply (i.e. > 60°), and so the throw of a fault is less than the amount of actual displacement along the fault plane itself.

On geological maps it is the convention to mark the downthrow side of a fault by a small tick mark, as shown in the symbols key. Most faults on the Moreton-in-Marsh map have the downthrow side shown.

❑   Look at the short fault trending north-west to south-east just south-west of Guiting Wood (SP 0626). Which is the downthrow side?

■   The tick mark shows that the downthrow is on the south-west side of the fault.

However, if in this case the downthrow had not been marked with a tick, you should still have been able to determine which was the downthrow side because the younger rocks would be exposed on the side downthrown by the fault (Section 2.4.3). This leads to a simple and reliable rule to apply in these cases:

Where younger rocks lie against older rocks along a fault displaying vertical displacement, the younger rocks occur on the downthrow side.

In the Moreton-in-Marsh area, the strata are virtually horizontal, so this rule can be easily demonstrated. Look again at the fault near Guiting Wood (Figure 3.6). Near its south-eastern end, the Fuller's Earth (FE) and Great Oolite Limestone (GOL) lie on the south-western side where they are faulted against the older Inferior Oolite (InO). At the north-western end of this fault, the Inferior Oolite (InO) lies on the south-western side and is faulted against the older Cotteswold Sands (Csd). Therefore, since younger strata occur on the south-west all along the fault, the south-west must be the downthrown side. However, some faults on the map do not have the downthrow side indicated.

Question 3.1   Examine the ENE to WSW-trending fault near Woolstone (SO 9630). This is an example where the map makers have chosen not to indicate the downthrow with a tick mark. Nevertheless, it is possible to determine downthrow from the outcrop pattern. Explain which is the downthrow side of this fault.

Most fault movements are intermittent and occur as periodic 'jumps' of a few centimetres or, at most, a few metres. Each jump causes a displacement of the rocks on either side of the fault plane and if the movement breaks the surface it can form a **fault scarp** (Figure 3.5a). During quiescent periods between the jumps, there is time for erosion to wear down the irregularities of the ground surface caused by the previous movement. Since the strata on the **upthrow** side of the fault are more elevated, they are eroded more rapidly, thus exposing the older strata beneath at a similar topographic level to the younger strata exposed on the downthrow side (Figure 3.7).

Figure 3.5   (a) Block diagram of a normal fault in horizontal strata; the downthrow side is on the left. (b) Cross-section to show the vertical displacement (throw) of strata by a normal fault. Note that because the fault is inclined, the total amount of movement along the fault plane is greater than that of the throw. (c) Sketch of relationship between displacement and throw (not to scale).

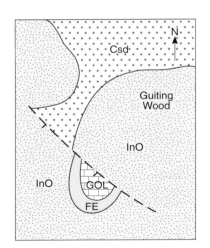

Figure 3.6   Sketch map of the fault at Guiting Wood. (The two tick marks show the downthrow side.)

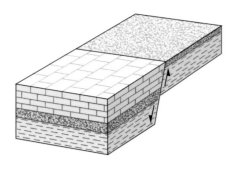

Figure 3.7   Outcrop pattern of the faulted strata shown in Figure 3.5(a) after erosion has worn down and levelled the elevated land surface behind the fault scarp.

Figure 3.8   (a) Block diagram of Hornsleasow graben before erosion. (b) Cross-section of Hornsleasow graben as it now appears. The strata are labelled as on the Moreton-in-Marsh map. (c) Block diagram of a horst before erosion. (d) Step faulting.

The combined effect of one or more faults can produce other structural features. On the Moreton-in-Marsh map, you have already observed that many faults have more or less parallel trends. Now look at the pair of generally WNW to ESE-trending faults to the north and south of Hornsleasow Farm on the map (SP 123322). Here, the downthrow of the adjacent faults is on opposite sides. The net effect is to produce a downfaulted block or **graben** between them. A graben is the structure formed when a block of country is *downthrown* between two parallel (or near-parallel) faults. Figure 3.8a and b illustrates this for the Hornsleasow area. The fault-bounded outlier you observed earlier at White Hall (SP 0122) could also be described as a graben.

A **horst** structure is formed when a block is *upthrown* relative to the surrounding country between two parallel (or near-parallel) faults, as shown in Figure 3.8c. A narrow horst block can be observed extending west from the exposure of Chipping Norton Limestone (CNL) at SP 110208 on the map.

Where faults are more or less parallel and have their downthrows on the same side, the resulting pattern is often called *step faulting* (Figure 3.8d).

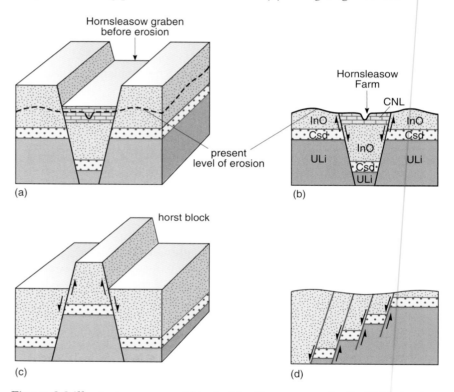

Figure 3.9 illustrates a second kind of fault, a **reverse fault**. This also has a steeply inclined fault plane, but the relative movement of the blocks on each side of this plane has the opposite sense to that of a normal fault. The Figure shows that one side has been pushed upwards *over* the other (cf. Figure 3.5 on the previous page).

The difference between normal and reverse faults is best summarized as follows:

For *normal* faults, the fault plane is either vertical or dips towards the *downthrown* block, while for *reverse* faults the fault plane dips towards the *upthrown* block.

Thrust faults are a special type of reverse fault. Here, the fault plane dips toward the upthrown block at a shallow angle (i.e. less than 45° from horizontal, but much less in many examples). You will learn more about these faults in Section 4.5.

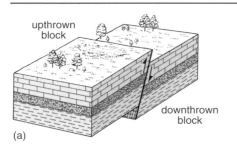 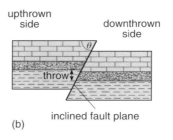 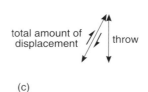

(a)

(b)

(c)

3.4.1 DETERMINING THE AGE AND THROW OF FAULTS

The relative age of a fault can be determined by establishing which stratigraphic units or geological features it cuts, and whether it has since been covered by later deposition. The following general rule applies:

> The age of a fault must be *younger* than the *youngest* bed (or any other geological feature) it cuts and *older* than the *oldest* bed (or any other geological feature) which is unaffected by the faulting. In other words, the youngest rocks or features that are cut by a fault give the *maximum* age of the faulting and the oldest rocks or features unaffected by a fault give the *minimum* age of faulting.

If we apply this rule to the faults of the Moreton-in-Marsh area, we can only make a very general statement as to the age of the faulting. Clearly, the faults cut all the Jurassic rocks but since there are no younger rocks covering the faults it is therefore not possible to determine a minimum age. All we can say is that faulting post-dates deposition of the Jurassic succession. Another important piece of information that can often be gleaned from geological maps is the amount of movement along the faults. This is often a matter of estimation rather than of precise measurement, but sometimes it is possible to estimate the throw at a particular locality to within a few metres.

Look now at SP 053246, where the upper and lower boundaries of Fuller's Earth (FE) have been faulted down on the south side of the fault to lie directly opposite the upper and lower boundaries of the Cotteswold sands (Csd). If we knew the exact thicknesses of the individual units at this locality, we could obtain a precise figure for the throw.

Question 3.2 Estimate the amount of throw on the fault at this point. (Assume that the Csd and FE are both 10 m thick here, that the Inferior Oolite (InO) is 50 m thick, and that the fault plane is vertical.)

In fact, we can estimate an upper and lower limit for the throw from the *range* of thicknesses in the generalized vertical section given for the Inferior Oolite (InO) and the Fuller's Earth (FE). Their combined maximum and minimum thicknesses can be used to obtain values for the maximum and minimum throw here.

Question 3.3 Use the maximum and minimum values given on the generalized vertical section for the thicknesses of the InO and FE beds to estimate maximum and minimum throws possible on this fault.

Many of the faults on the map have a finite length. If you look again at the fault near Guiting Wood (SP 0626), you can see that at each end the fault dies out. This can only mean the amount of throw along a fault must vary from zero at the end of the fault line up to some maximum amount somewhere in between (Figure 3.10).

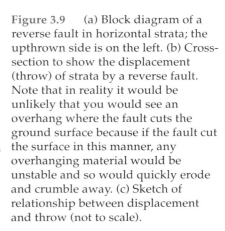

Figure 3.9 (a) Block diagram of a reverse fault in horizontal strata; the upthrown side is on the left. (b) Cross-section to show the displacement (throw) of strata by a reverse fault. Note that in reality it would be unlikely that you would see an overhang where the fault cuts the ground surface because if the fault cut the surface in this manner, any overhanging material would be unstable and so would quickly erode and crumble away. (c) Sketch of relationship between displacement and throw (not to scale).

Activity 3.1

You should now do Activity 3.1. The first part is designed to aid your understanding of how movement on faults and fault structures affects stratigraphy. The second part is a 'field trip' to Big Bend National Park (USA) where you will have a chance to look at a landscape produced by large-scale faulting.

Figure 3.10 Block diagram illustrating the way in which a fault can die out at each end. The sag in the middle has been exaggerated for the purposes of this Figure and need not be symmetrical. The inset block is Figure 3.5a.

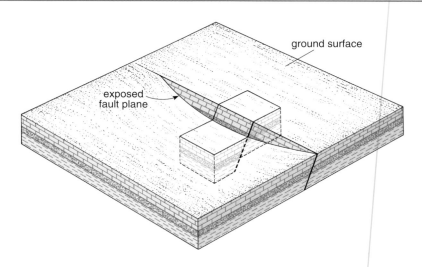

ground surface

exposed fault plane

3.5 CONSTRUCTING AN ACCURATE GEOLOGICAL CROSS-SECTION

You have already been guided through the construction of a simple map and a cross-section (Activities 2.2 and 3.1). You should now be able to construct cross-sections across areas of the Moreton-in-Marsh Sheet. Whenever you construct an accurate cross-section on paper, you should use the guidelines given in Box 3.1.

Box 3.1 Construction of an accurate cross-section

• Your section should be at least as deep as sea-level, and usually below sea-level (the depth to which you can usefully extend cross-sections downward is usually greater in areas of more complex tilted, folded and faulted rocks).

• If possible, keep vertical and horizontal scales similar, especially if you want to use the dip information (Section 3.2.8). However, on more complicated sections, this may not be practical, and in order to show subsurface detail, some vertical exaggeration may be necessary.

• Draw the topographic profile first. Label the ends of the section with a grid reference or compass bearing. In the most modern maps, it is conventional to put the westernmost end of the section at the left, and for exactly north–south sections to put south on the left. You can adopt your own convention, but be *consistent*.

• Transfer the points of outcrop of boundaries of stratigraphic units, faults and other geological features from the map to the topographic profile. This can be achieved by using a piece of graph paper laid along the chosen line of section and then marking where the geological boundaries or faults occur. If you lay your marked graph paper along the

topographic profile, you can then mark precisely where they occur on the topography.

• Choose an important stratigraphic boundary and draw this across the whole section, joining up the points where it intersects the topography. This will then provide a marker horizon which will help to construct the overall structure of the beds. Put in other units in the same series of strata by determining their outcrop positions from their intersections with the topographic profile and then drawing them as lines parallel to the first one. Try to ensure that the thicknesses of the individual beds you have drawn lie within the range indicated in the stratigraphic column and, if it is drawn without vertical exaggeration, that they remain constant across your section.

• Determine the downthrow side of each fault. Assume faults are normal faults unless there is clear evidence to the contrary, and draw them dipping steeply toward the downthrow side. Where possible determine the amount of throw on the fault and show this on the section. Label the direction of displacement by using half-arrows

• If there is insufficient information for you to be sure of boundaries at depth, on part of the cross-section indicate this by use of a question mark.

• Prepare a labelled key by colour shading and/or lettering the beds shown on the section.

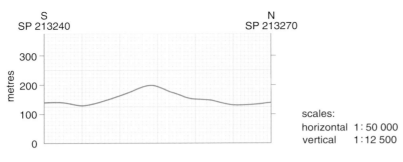

Figure 3.11 Topographic profile between SP 213240 and SP 213270 for use with Question 3.4. Vertical exaggeration is × 4.

Question 3.4 Construct an accurate cross-section from SP 213240 to SP 213270 using the topographic profile provided in Figure 3.11. Horizontal scale is the same as the map, but a vertical exaggeration of × 4 has been used to enable you to show the relationships of the rocks more easily. For the purposes of this exercise, assume the beds are horizontal (arrows indicate only very shallow dips here), and ignore the superficial Boulder Clay deposits.

Question 3.5 (a) Construct an accurate cross-section from SP 055340 to SP 110315 using the topographic profile provided in Figure 3.12.

(b) How would you describe the region composed of MLi through to CNL inclusive upon which the settlements of Stow-on-the-Wold and Maugersbury (SP 1925) have been constructed?

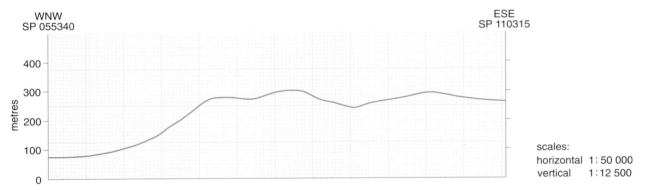

3.6 DETERMINING A GEOLOGICAL HISTORY FROM THE MAP

Figure 3.12 Topographic profile between SP 055340 and SP 110315, for use with Question 3.5. Vertical exaggeration is × 4.

A geological history is simply a sequence of events. The timing of different geological events (e.g. which beds were deposited when, and whether they were later faulted or not) can be determined from the type of outcrop patterns we have already discussed.

We can already say that because there are no large stratigraphic gaps (unconformities) within the solid geology exposed in the area of the map (i.e. Lower Lias to Forest Marble), they must have been laid down in a **conformable succession** during the Jurassic, and then at some later time became faulted, after which uplift and erosion have led to the development of the present-day topography. For large-scale maps such as the Moreton-in-Marsh Sheet, which cover small geographical areas and a small part of the stratigraphic column, the geological history often just consists of a few such clearly defined local events involving deposition of sediments and their subsequent folding and/or faulting. However, for small-scale maps such as the Ten Mile Map, unravelling the geological history becomes much more complex because more geological time and a much greater area is involved.

Activity 3.2

You should now do Activity 3.2, which is a guided exercise in compiling a geological history, using the Moreton-in-Marsh area as an example.

3.7 CHECKLIST: HOW TO LOOK AT IGS/BGS 1:50 000 GEOLOGICAL MAP SHEETS OF THE UK

Most of the modern IGS/BGS geological map sheets use a similar format to that of the Moreton-in-Marsh Sheet. The following represents a summary of the observations and interpretations outlined in Section 3 and offers guidelines as to how you might tackle similar map sheets. (*Note:* non-British geological maps often have a different format.)

1 Locate the sheet on the Ten Mile Map (the edges of all One-Inch or 1:50 000 maps are overprinted in red on the Ten Mile Map together with the sheet numbers).

2 Determine the scale of the map.

3 *Geographical Information* of the base map (usually printed in grey):
   (a) Determine the nature of the grid and spacing of grid lines;
   (b) Are contours in feet or metres, and what is the contour interval?;
   (c) When was the geographical survey carried out (i.e. is it relatively modern)?

4 *Geological Information* (usually printed in black):
   (a) Which edition? Solid, Drift, or Solid and Drift?
   (b) Key (often called 'explanation of geological symbols and colours'). Establish which rock types and ages are present from the stratigraphic column; solid and drift information is usually separate. Are there any special symbols given, such as overturned or vertical strata, or thrusts?
   (c) *Generalized vertical section*(s) – Observe the variations in thicknesses of strata which occur within the area of the map. Note the scale(s) of these vertical columns and the area of the map to which a specific column applies. Note especially the presence of unconformities and if wedging out of strata occurs.
   (d) *Cross-section*(s) – Determine if any vertical exaggeration has been applied. Locate where the line(s) of section cross(es) the map.
   (e) If given, what is the history of the geological survey(s) upon which the map is based? Read any note regarding the geological history of the area.

3.8 OBJECTIVES FOR SECTION 3

Now you have completed this Section, you should be able to:

3.1   Use the information provided in the margins of a geological map sheet to interpret the geological features shown on the map.

3.2   Recognize the evidence for horizontal or gently inclined strata from the outcrop patterns.

3.3   Distinguish between outliers, inliers and areas of superficial drift deposits.

3.4   Identify strata of variable thickness on the generalized vertical section or the map.

3.5   Recognize faulting and the type of outcrop patterns resulting from normal faults.

3.6   Determine the relative age of a fault from map evidence.

3.7   Determine the downthrow side of a normal fault using stratigraphic evidence.

3.8   Recognize the presence of horsts and grabens from the pattern and relationships of the faulting.

3.9   Measure or estimate the amount of throw on a normal fault which displaces horizontal strata.

3.10   Draw accurate cross-sections across key areas of a geological map depicting an area of horizontal or gently inclined strata.

3.11   Identify unconformities in the generalized vertical section, and be able to give a simple geological history of how they may have formed.

3.12   Utilizing all the relevant map sheet information, compile a simple geological history of the area.

Now try the following questions to test your understanding of Section 3. Questions 3.6–3.8 use the Moreton-in-Marsh Sheet, and Questions 3.9–3.12 use the Ten Mile Maps.

> **Question 3.6**   (a) Complete the following statement: Chipping Campden (SP 1538) is built largely on strata of .... age.
>
>
>
> (b) Delete the incorrect words in the following statement:
>
> The road from Chipping Campden which runs south, then south-west toward Lapstone Farm (SP 142364) *climbs/descends* over *younger/older* strata than those under Chipping Campden.
>
> (c) What lines of evidence suggest that the strata in this part of the map are horizontal?
>
> **Question 3.7**   (a) Why can the hills of MLi and ULi at SP 0035 and SP 0034 south of Dumbleton be correctly described as outliers?
>
>
>
> (b) Why would it be incorrect to describe the outcrop of Cheltenham Sands at Alderton (SP 001334) as an outlier?
>
> (c) How would you describe the type of *structure* comprising the FE and GOL succession developed between the two faults in the tract in which White Hall lies (SP 0122)?
>
> **Question 3.8**   Find the fault trending WNW to ESE at SP 0135 south of Dumbleton.
>
>
>
> (a) Which is the downthrow side of the fault, and why?
>
> (b) Estimate the amount of throw on the fault at (i) SP 013351 and (ii) SO 998356.
>
> (c) Locate the N–S-trending normal fault extending between SP 122304 and 123287 east of Trafalgar Farm and determine the downthrow side.
>
> **Question 3.9**   Look at the fault trending NE to SW through Wem (SJ (33) 5128) on the Ten Mile Map (S).
>
>
>
> (a) Which is the downthrow side of the fault, and why?
>
> (b) Why is it not possible to determine the amount of throw on this fault?
>
> (c) Are the Middle Lias rocks (92) near here outliers or inliers?

Question 3.10 Using the stratigraphic information given on the Ten Mile Map (S), estimate a minimum and maximum age for movement along the Bala fault (SH (23) 5800–SJ (33) 3357).

 Question 3.11 (a) On the Ten Mile Map (N, *not* S!), is the pale blue outcrop (80) which extends NW from Scotch Corner (NZ (45) 2206) an outlier or an inlier?

(b) How would you describe the tiny outcrop of Carboniferous Coal Measures rocks (82–3) at NZ (45) 0732 lying WNW of Hamsterley (NZ (45) 1131)?

 Question 3.12 Using the Ten Mile Map (S), on a sheet of paper sketch a cross-section from Gloucester (SO (32) 8317) to Marlborough (SU (41) 1868). Put the north-western part of the section on the left. (*Note:* Assume flat topography and do not worry about accuracy since you have no direct evidence as to the amount of dip. Show only the thicker units on your sketch section.)

4 THE CHEDDAR SHEET, ST 45

 Now you are familiar with geological map outcrop patterns formed in areas of horizontal strata, we can turn to those representing inclined and folded sediments. In order to understand such features, you will need to be able to understand fully the principles of strike and dip. Your knowledge of fault interpretation will also be extended so you can determine the nature and degree of relative displacement along faults where the strata have been tilted and folded.

 For most of this Section, you will be using the Cheddar Sheet from your Home Kit. You will also need your Ten Mile Map (S), a protractor and Multimedia access.

4.1 INTRODUCTION

Most of the remaining mapwork principles with which you need to become familiar are superbly illustrated by the Cheddar Sheet. There is a major unconformity on this map, and this will be used to show you how to identify this type of important feature. Other key objectives are to learn how to recognize folding from the outcrop patterns, and how to draw cross-sections across areas where the strata are folded and where unconformities occur. When you have finished studying this sheet, you should be able to understand almost any geological map of sedimentary strata.

4.1.1 GEOGRAPHICAL INFORMATION

First, locate the position of the sheet on the Ten Mile Map (S); it lies SSW of Bristol. The sheet number ST 45 is the reference of the 10 km grid square which this sheet depicts, namely ST (31) 45. The northern half of the Cheddar Sheet covers part of the western end of the Mendip Hills in Somerset; the town of Cheddar is just south-east of the middle of the map, and the famous Cheddar Gorge can be seen just to the east of the town (ST 4754). The Mendips lie to the west of the main Jurassic escarpment (Section 3.2), which forms part of the northern extension of which you have studied on the Moreton-in-Marsh Sheet.

Before looking at the Cheddar map itself, use the preliminary approach outlined for the Moreton-in-Marsh Sheet and familiarize yourself with *all* the information in the margins. Notice in particular the differences from the Moreton-in-Marsh

Sheet: (i) the Cheddar map is at twice the scale of the Moreton-in-Marsh map, i.e. 1 : 25 000, and (ii) the contours are in feet, not metres.

Activity 4.1

To help you assimilate the key background information, now complete the checklist in Activity 4.1.

4.1.2 BACKGROUND GEOLOGICAL INFORMATION

The rocks exposed at the surface on the Cheddar Sheet represent a much longer span of geological history than those exposed in the Moreton-in-Marsh area. The Mendip Hills are formed of folded Devonian and Carboniferous strata which were subsequently eroded and then covered by Triassic and Jurassic sediments.

The Devonian rocks (orange-brown) and Carboniferous rocks (blues and greens) crop out in a series of sub-parallel belts running east–west across the northern half and south-eastern quarter of the map. In the north, west, and south of the map, Triassic Keuper Marl (Mercia Mudstone Group) sediments (light pink) partially obscure the outcrop patterns of these earlier Carboniferous rocks. The youngest rocks of the stratigraphic succession of the Cheddar area are Lower Lias of Jurassic age (grey and light brown); they are equivalent in age with the oldest rocks exposed in the Moreton-in-Marsh area. Outcrops of Lower Lias occur in the south-west corner and are separated from the Mendip Hills by a north-west to south-east-trending belt of estuarine alluvium and silty river channel material (pale yellows and light browns) which together constitute the drift deposits in this area.

4.1.3 THE GENERALIZED VERTICAL SECTION

The generalized vertical section for the solid geology is drawn to scale and therefore gives a direct visual guide to the range of thicknesses of the stratigraphic units in this area. Within the area of the map, there is considerable variation in the thickness of individual beds as is demonstrated by the thickness ranges given alongside most of the stratigraphic units. In addition, some units also have brief lithological descriptions. Some of the terms used here may be unfamiliar to you:

conglomerate: a sedimentary deposit typically consisting of large rounded or sub-rounded gravel-, pebble-, or cobble-sized fragments (i.e. 2 to >256 mm), often of variable lithology, and commonly set within finer-grained (e.g. sandy or muddy) materials;

marl: a mudstone with a large calcium carbonate content;

crinoidal limestone: limestone very rich in remains of crinoids (a type of marine animal);

chert: a form of very finely crystalline silica (similar to flint);

dolomite: a carbonate mineral with the formula $CaMg(CO_3)_2$. In this case, the term refers to a *rock* which contains a considerable amount of this mineral;

mineral vein: a fissure or crack in a rock now filled with one or more minerals.

In the older styles of IGS map, the strata are identified by a combined letter and number system rather than the letter-only abbreviations you saw used on the Moreton-in-Marsh map. You have already been introduced to the numerical system on the Ten Mile Map (Section 1.5). Table 4.1 shows how the older letter and number system relates to that of the Ten Mile Map:

1111

Table 4.1 Relationships between the numbering convention on the Ten Mile Map and the letter system based on older IGS maps. (*Note:* you are not expected to memorize these.)

Era or Period	No. on Ten Mile Map	Letter on older IGS maps
Quaternary	115	l
Tertiary	107–114	i, j, k
Cretaceous	102–106	h
Jurassic	91–101	g
Triassic	90	f
Permian	85–89	e
Carboniferous	79–84	d
Devonian	75–78	c
Ordovician–Silurian	67–74	b
Cambrian	62–66	a

Each letter of the older IGS system referred to a period of the geological column. For instance, the Carboniferous rocks, such as those of the Cheddar Sheet, were designated 'd'. To differentiate different stratigraphic units within the Carboniferous succession, a superscript number and letter was added to the letters to represent further subdivisions. The units were designated in order of decreasing age within each lettered subdivision (i.e. d^1 is older than d^2, and d^{1a} is older than d^{1b}). Stratigraphic units on your Cheddar Sheet are identified according to one or other of these conventions depending on whether you have a newer or older edition of the map. A comparison of the two systems (i.e. letter–number and an abbreviation convention) for the Devonian, Carboniferous and Triassic strata of the Cheddar Sheet are given in Table 4.2. (*Note:* you are not expected to learn these.)

Table 4.2 A comparison of stratigraphic names and abbreviations on older and newer versions of the Cheddar Sheet, showing their equivalents on the Ten Mile Map. (*Note:* the Quartzitic Sandstone Group does not outcrop within this sheet, but is inferred in the cross-section.)

Geological Period	Stratigraphic unit	Abbreviation system (BGS maps)	Letter-number system (old IGS maps)	No. on Ten Mile Map
Triassic	Mercia Mudstone Group (or Keuper Marl)	MMG	f^6	90
Carboniferous	Quartzitic Sandstone Group	QSG	d^4	81
	Hotwells Limestone	HL	d^3	80
	Clifton Down Limestone	CDL	d^2	80
	Cheddar Limestone	CdL	d^2	80
	Cheddar Oolite	CO	d^2	80
	Burrington Oolite	BO	d^2	80
	Black Rock Limestone	BRL	d^{1b}	80
	Lower Limestone Shale	LSh	d^{1a}	80
Devonian	Portishead Beds	PoB	c^3	78

A very important stratigraphic feature at the bottom of the Keuper Marl (or Mercia Mudstone Group – MMG) is a major unconformity which is shown by a wavy line on the generalized vertical section. You met this concept briefly when studying the Ingleton area (Section 2.4.2) and the Moreton-in-Marsh vertical section (Section 3.2.5). In the Cheddar area, much of the Upper Carboniferous and the whole of the Permian are absent, so this is a major unconformity representing a gap from about 300 Ma to about 230 Ma (i.e. ~ 70 Ma in duration).

4.1.4 Wedging-out of strata and lateral changes in the nature of beds

On the Moreton-in-Marsh Sheet, you saw how some units in the stratigraphic succession thinned out and disappeared in some areas of the map (Section 3.3.4). On the Cheddar Sheet, several units of the Carboniferous Limestone Series in the stratigraphic column are 'invaded' by different-coloured wedge-shaped and rectangular features, some of which have been given separate names. These indicate the stratigraphic position of rock units that are present over only part of the map. Another feature of the sedimentary succession in this area is that the composition (i.e. lithological characteristics) of a particular rock unit or bed can change laterally along its outcrop. Such lateral changes are typically the result of different depositional conditions having existed at the same time. Since the resulting different lithologies are of the same age, they can be correlated (Section 1.5) and so need to be shown occupying a similar position in the vertical section. For example, at the base of the Lower Limestone Shale (i.e. LSh or d^{1a} – see Table 4.2) there is a 'limestone wedge' in the column indicating that a limestone was being deposited at some localities within the Cheddar area at the same time as shales (i.e. silty or muddy sediments) were being laid down elsewhere.

4.1.5 Drift deposits

As in the Moreton-in-Marsh area, the drift deposits have a patchy distribution and their thickness is not shown to scale in the map key. A few terms may be unfamiliar:

Head: a term given to typically chaotic deposits formed around a glaciated area and resulting from waterlogged soil flowing downslope when the ground began to thaw;

calcareous tufa: a precipitate of calcium carbonate limestone deposited by springs in a limestone area;

terrace deposit: a level of a former flood plain of a river, abandoned as the river cuts deeper into its valley;

alluvial fan (or *alluvial cone*): a fan-shaped deposit of sediments produced when a river dumps its load as its velocity is reduced (e.g. upon emerging from a steep valley onto a flat plain).

4.2 The cross-section

The map scale is large enough for the beds to be easily represented on the section at true scale, so both vertical and horizontal scales are at 1 : 25 000, thus avoiding vertical exaggeration. This is advantageous because ground surface slopes, dip angles of the strata and their stratigraphic thicknesses are shown with their true values. The section runs in a dog-leg from the north-east to south-west corner of the map. In the south-western part of the cross-section, the beds are almost horizontal whilst the north-eastern part shows that the strata form a large anticlinal fold. North of Black Down, the Carboniferous strata dip steeply to the north while to the south they dip more moderately southwards.

❑ Determine the amount of dip of the strata to the north and south of Black
Down from the cross-section. Remember that dip is always measured from
the horizontal and, because there is no vertical exaggeration, you can
measure it directly with a protractor. Compare your answer with the dip
arrows marked on the map.

■ To the north of Black Down, the cross-section indicates a dip of about 70°–
80°, whilst on the map dip values range from about 45°–64° near the line of
section. By contrast, south from Black Down to Cheddar Gorge the dip
measured from the cross-section is only in the region of 25° and is similar to
the dip values of 20°–27° marked on the map.

4.3 THE MAP

The aim of this Section is to investigate the outcrop patterns exhibited by
dipping strata. We shall therefore concentrate upon the elongated outcrops of
Carboniferous strata (Lower Limestone Shale to Hotwells Limestone) flanking
the area of Old Red Sandstone strata (Portishead Beds) which together define
the anticlinal structure.

4.3.1 STRIKE AND DIP OF BEDS

Where beds have been tilted and folded by earth movements, as in the Cheddar
area, the dip symbols provide important clues to geological structure. From the
Moreton-in-Marsh map, you should remember that horizontal strata are
represented by a cross symbol and dipping beds by arrows showing the *direction*
in which the strata are inclined. The amount of dip in degrees from the
horizontal is often printed beside the arrow. The attitude of inclined strata is
always described in terms of their *strike* and *dip* (Figure 4.1): these terms were
introduced and explained in Section 2.3.1. However, to interpret patterns of
folding, we first need to understand more fully the principles of strike and dip.

The **strike** is the *direction* of a horizontal line on an inclined surface (e.g. an
inclined bedding plane). The direction of strike is measured from the true
north and is often calculated from a compass bearing. For example, in Figure
4.1 the beds have a N–S (or S–N) strike which can be expressed as a strike of
either 000° or 180°.

The **dip** is the angle of maximum slope of an inclined surface. It has two
components: (i) an *amount of slope* (inclination), expressed as an angle in
degrees from the horizontal (or more rarely as a gradient); and (ii) a *direction*,
which is *always* perpendicular (i.e. 090°) to the strike and is in the *down*slope
direction. On Figure 4.1, the amount of dip is 60° towards the east (090°).

There are two conventions used to show strike and dip on geological maps. On
BGS maps, dip directions are usually shown by arrows with the amount of dip
in degrees indicated alongside. The strike direction is at right angles to the dip

Figure 4.1   Block diagram of tilted
strata illustrating the principles of
strike and dip. The symbols show the
two conventions used to indicate
dipping beds on geological maps.
(Note that this Figure has been
repeated from Figure 2.5 for
convenience.)

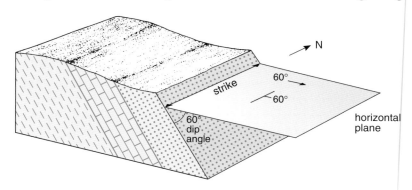

direction. However, sometimes a line is drawn to show the strike direction, with a small tick bar to indicate dip direction and a number to indicate the amount of dip. In Figure 4.1, these two types of symbols are shown as they would appear on a map. Look again at Box 2.1 if you want to remind yourself of how strike and dip are recorded.

Whenever the term dip is used without qualification, it is always **true dip** (i.e. the *maximum* inclination of that surface) that is implied. An **apparent dip** is the amount and direction of dip measured on the dipping surface in any direction other than true dip. Apparent dips are always *less* than true dip; they are *not* shown on geological maps.

A simple analogy is to consider one side of the sloping roof of a house (Figure 4.2). The ridge of the roof lies along the strike direction of the roof surface, whilst the dip direction and amount can be measured along the gable end of the roof since these are at 90° to the ridge of the roof (i.e. strike). Any other line drawn across the roof will not be at 90° to the ridge and so its dip will be an *apparent* dip and will be *less* than that of the *true* dip. As shown in Figure 4.2, there are any number of apparent dips varying from very shallow angles (*a*) nearly parallel to the strike direction to angles almost as steep (*d*) as the (true) dip.

Strike and dip define *uniquely* the orientation of any planar surface. In geology the measured surface is commonly a bedding plane, but it can also be a fault plane, a mineral vein or any other geologically important planar feature. The information conveyed in strike and dip measurements is crucial in understanding structures in areas of faulting and folding.

4.3.2 STRIKE ON GEOLOGICAL MAPS

From Section 2.3.1, you should remember that in areas of dipping strata the trend of the outcrop pattern at the surface gives a good indication of general direction of strike (unless the topography is sufficient to affect significantly the shape of the outcrop pattern, which can only occur where dips are gentle). Now look at the northern half of the Cheddar Sheet at the blue and green strips representing the outcrops of Carboniferous rocks (Lower Limestone Shale to Hotwells Limestone) and the orange–brown strip of Old Red Sandstone strata (Portishead Beds).

❏ What is the relationship between the dip arrows and the pattern of outcrop of these beds?

■ The dip arrows are always at 90° (or nearly so) to the boundaries between the beds. In other words, the strip pattern of the different units is parallel to the strike of both the Carboniferous and Old Red Sandstone rocks of this region.

Once you have established the general strike, it becomes much easier to work out the dip direction even in areas where few dip arrows are marked on the map: since the northern half of the Cheddar map has a dominant east–west strike, the dips must be *either* to the north *or* to the south.

4.3.3 DETERMINING DIPS FROM STRATIGRAPHY

It is relatively straightforward to make interpretations of geological structure if plenty of dip arrows are provided. But suppose the map had neither dip arrows nor a cross-section, and the only information were the outcrop patterns on the map and the stratigraphic column. How, for instance, would you be able to determine the direction of dip of the Carboniferous succession (Limestone Shale–Hotwells Limestone) and Old Red Sandstone (Portishead Beds)? If you know the stratigraphic succession, then this is quite easy because in a succession

❏ What would a value of 020/35E measured on a bedding plane signify in a notebook?

■ It would indicate a strike of 020° and a bedding dip of 35° in an easterly direction.

In this case the dip is not exactly east. That does not matter because 020 is the precise reading and E merely serves to indicate the dip is toward the east, rather than toward the west. Moreover, since we know dip and strike are at right angles to each other, it follows that the precise dip and dip direction of the bed is 35° toward 110° (i.e. ESE).

Figure 4.2   House roof analogy to illustrate dip and strike. True dip is parallel to the gable end, in other words at 90° to the roof ridge. Strike direction is parallel to the roof ridge. Apparent dips *a*, *b*, *c* and *d* are also shown.

Figure 4.3   Block diagram illustrating that progressively younger strata are encountered in the direction of dip. Top surface of the block represents the outcrop pattern as seen on a map.

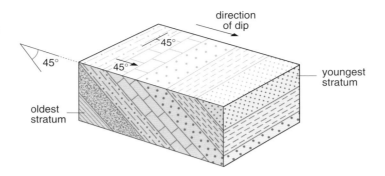

of tilted strata progressively younger beds crop out in the direction in which the strata dip (Figure 4.3). This leads to the following general rule:

Younger strata crop out in the direction of dip.

Now look at the northernmost E–W-striking strips of Carboniferous strata on the map.

❑   Determine in which direction they become younger.

■   They become younger toward the north with the oldest Carboniferous beds (Limestone Shale) and Old Red Sandstone (Portishead Beds) forming the core of the anticlinal structure. Therefore the dip direction of these Carboniferous strata is northwards on the northern limb of the anticline.

4.3.4   DETERMINING DIPS FROM OUTCROP V-PATTERNS

Now let us suppose that you do not even know the stratigraphic succession of the rocks! How could you then tell the direction of dip? We are now only left with the outcrop pattern on the map, but remember that in Section 3.3.2 you were shown how outcrop V-patterns in valleys helped determine areas of horizontal strata. We shall now explore how V-patterns also relate to valleys cut into areas of dipping strata since these can be often used to deduce the dip direction of inclined strata, and in some cases to estimate whether the dip is steep or shallow. Figure 4.4 shows the two cases for strata dipping in opposite directions across a river valley and illustrates a very useful rule:

Outcrop Vs of *dipping strata* on geological maps typically point in the direction of dip.

❑   Look carefully along the southern strip of Black Rock Limestone between easting grid lines 44 and 46. Do you see any outcrop V-patterns?

■   There is a clear V-pattern which cuts across the topographic contours at ST 446564 along the boundary between the Lower Limestone Shale and Black Rock Limestone, and south of Batt's Farm (ST 451556) between the Black Rock Limestone and Burrington Oolite. These V-patterns point south, so the strata must also dip in this direction. The V-pattern can also be seen in the chert bands in the Black Rock Limestone. The map dip arrows confirm a southerly dip of 25°–30° in this area.

Activity 4.2

Now do Activity 4.2. This is designed to aid your understanding of how dipping strata affect outcrop patterns.

The other important piece of information which can often be determined from V-pattern relationships is the steepness of dip. From Figure 4.4 b and c we can see that for a valley of given topographic slope the outcrop V-pattern becomes less pronounced as the dip of the strata steepens.

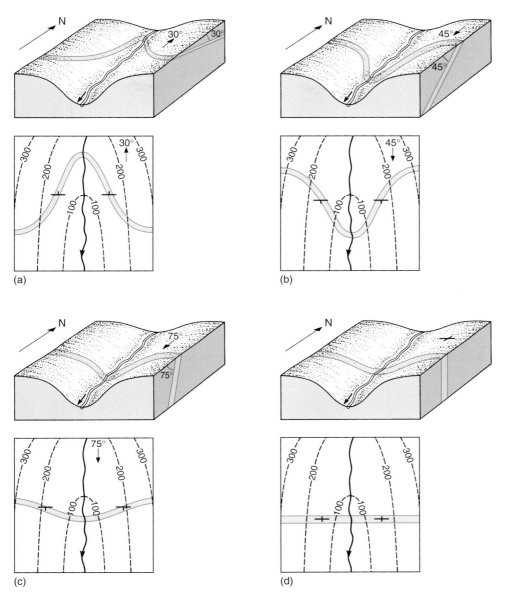

Figure 4.4 Block diagrams and associated plan views (i.e. map view) illustrating outcrop Vs on inclined beds.
(a) Bed dips northwards, upstream, so producing a V-pattern pointing upstream. Note also that because the beds dip they cross the topographic contour lines (shown as dashed lines).
(b) Beds dip southwards, downstream (about 45°) and, because they dip, they also cross the contour lines so producing a V-pattern pointing downstream.
(c) Beds dip southwards as in (b), but here the dip is much steeper (about 75°): in the plan view the V-pattern still points downstream, but is much less pronounced.
(d) Vertically dipping strata showing no V-ing of the outcrop pattern as it crosses the valley.

❑ Look carefully along the eastern end of the northern strip of Old Red Sandstone (Portishead Beds) and Carboniferous strata (Lower Limestone Shale–Clifton Down Limestone) near Middle Ellick Farm (ST 494575). Can you see any outcrop V-patterns?

◼ V-patterns certainly occur in only two or three places in the stream valley, but are not so noticeable as those you saw earlier on the map. They are best seen along the boundary between the Portishead Beds and the Lower Limestone Shale, around ST 493574, and in some of the chert bands in the Black Rock Limestone (e.g. at ST 493578). Nevertheless, because the strata show a V-shape pointing north, this must be the direction of dip.

The map dip arrows confirm the northerly dip predicted by the V-pattern, but you should also note that the dip of the beds exposed in this stream valley is quite steep, 45°–55°, compared with 30° near Batt's Farm. Moreover, the stream valley rising near Middle Ellick Farm continues to the north, through the Black Rock Limestone and Burrington Oolite–Clifton Down Limestone succession, but there are no obvious V-patterns here. Inspection of the dip values within these beds reveals that they are very steep, up to 70°–80°, and thus are approaching vertical.

❏ What type of outcrop pattern do you think might occur if the beds crossing the river valley in Figure 4.4b were dipping vertically?

■ As the dip of strata becomes steeper, the V-pattern becomes less and less to a point where, in the case of very steeply dipping beds (typically 75° or more), valley V-patterns are no longer obvious. The outcrop relationship for vertically dipping strata is illustrated in Figure 4.4d.

This relationship presents another important rule:

A vertical stratum, or any other vertical geological feature, does not show an outcrop V-pattern as it crosses a valley. Instead, its outcrop will be straight.

Note that the rules about V-ing we have been discussing also apply equally well to other vertical planar features, such as fault planes or sheets of intrusive igneous rocks such as dykes.

4.3.5 STRATUM THICKNESS AND WIDTH OF OUTCROP

Another consequence of variation in the dip of strata is differences in the width of outcrop (Section 2.3.3). Remember that the relationship between topographic slope and outcrop width was illustrated for horizontal strata in the Moreton-in-Marsh area (Section 3.3.3). Now look again at the Carboniferous strata on the northern and southern limbs of the anticline on the Cheddar map. The width of outcrop of the Carboniferous strata (e.g. Lower Limestone Shale and Black Rock Limestone) in the southern strip is about twice that of the same strata in the northern strip, but examination of the topographic contours shows that the slopes in the two areas are broadly similar. Therefore, unlike the outcrop patterns of the Moreton-in-Marsh map, there is no clear relationship between outcrop width and topography. You have already made the key observation that dip in the northern strip is much greater than that of the southern strip and, in fact, the differences in outcrop width are the result of this variation in dip of the strata. Therefore, there are two main factors which control outcrop width: (i) the steepness and direction of topographic slope; and (ii) the amount of dip.

We will now look at the relationship between outcrop width and dip in detail, assuming an essentially horizontal ground surface. Figure 4.5 shows that for a stratum of given thickness, the steeper the dip (θ_1), the narrower the width of outcrop (w_1), and the shallower the dip (θ_2) the wider the outcrop (w_2). Moreover, the **true thickness of the bed** (t), the angle of dip from the horizontal (θ), and the outcrop width (w), can be related by the following simple formula:

$$\sin \theta = t/w \text{ (see Figure 4.6)} \tag{4.1}$$

Therefore, rearranging the equation:

true thickness (t) = width of outcrop (w) × sine of angle of dip (θ)

$$t = w \sin \theta \tag{4.2}$$

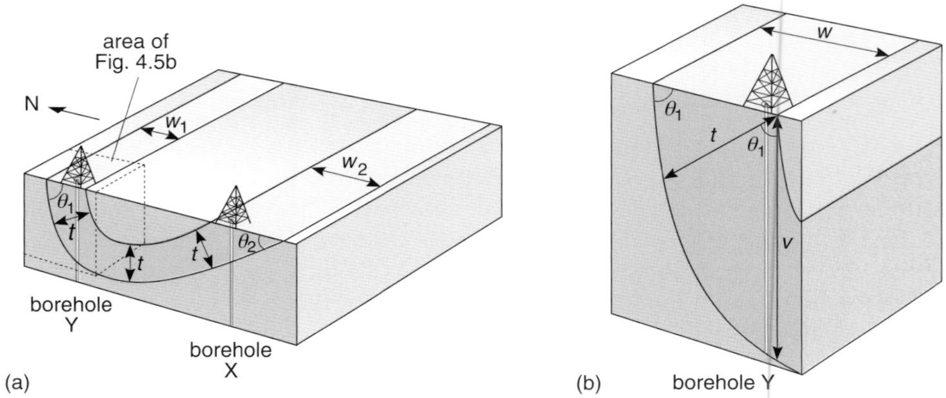

Figure 4.5 (a) Block diagram showing a bed of constant thickness (t) folded into an asymmetrical syncline (i.e. the northern limb dips more steeply than the southern limb). As a consequence of the different dip angles, the width of outcrop (w_1) is less than w_2 because θ_1 is much steeper than θ_2. Moreover, a vertical borehole at X would go through much *less* of a vertical thickness of the folded bed than a similar borehole at Y. (b) Enlargement of NW corner at the location of borehole Y *to* illustrate the relationship between dip (θ), width of outcrop (w), true thickness (t) and vertical thickness (v).

Although this formula is strictly accurate only when the topography is flat, it is nevertheless often useful in helping determine the approximate thickness of beds from geological maps. For most examples in this Course, this is valid because the outcrop widths are generally large compared with the topographic relief.

> **Question 4.1** Calculate the approximate true thickness of the Black Rock Limestone in two places (assume that the topography is flat across the Black Rock Limestone outcrop in both places). You may get two different thickness values, but this is nothing to worry about:
>
> (a) near ST 462562, where there are two dip arrows of 28° and 29°; and
>
> (b) near ST 454586, south of Dolebury Warren, where rather greater dips occur (70° and 80°).

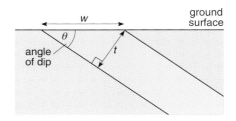

Figure 4.6 Line diagram showing the trigonometrical relationship between width of outcrop (w), angle of dip of bedding from the horizontal (θ), and the true thickness (t) of the bed.

Next, it is important to understand how measurements of the **vertical thickness of a bed** can vary in areas of folded rocks. Figure 4.5a shows that if we were to put down a vertical borehole at point X then we would drill through less of a vertical thickness of layer b than if we were to sink a similar borehole at Y. This relationship between vertical thickness (v), dip (θ) and width of outcrop (w) is illustrated in Figures 4.5b and 4.7, and gives the following equation:

$$\tan \theta = v / w \text{ (see Figure 4.7)} \tag{4.3}$$

Therefore,

$$\text{vertical thickness } (v) = \text{width of outcrop } (w) \times \text{tangent of angle of dip } (\theta)$$

$$v = w \tan \theta \tag{4.4}$$

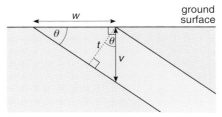

Figure 4.7 Line diagram showing the trigonometrical relationship between width of outcrop (w), angle of dip of bedding (θ) and vertical thickness (v) of the bed.

It follows that vertical thickness (v) and true thickness (t) are related by the formula

$$\cos \theta = t/v \tag{4.5}$$

Therefore:

$$v = t/\cos \theta \text{ (see Figure 4.8)} \tag{4.6}$$

Finally, in some cases, topographic slope can become important in controlling width of outcrop. For example, on the map cross-section, compare the width of outcrop of the Burrington Oolite on both sides of the anticline. To the south it has a very wide outcrop because it has a dip roughly similar to that of the topography (i.e. it forms a dip slope). The opposite is true in the north where it has a very narrow outcrop, with dips up to 80°. This leads us to another very important point:

> The width of outcrop of a vertical stratum is always equal to its true thickness, and so can be determined directly from the map. It is unaffected by topography.

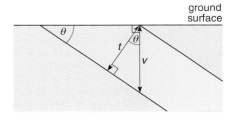

Figure 4.8 Line diagram showing the trigonometrical relationship between true (t) and vertical (v) thickness of the bed, and the angle of dip of bedding (θ).

4.4 FOLDS ON GEOLOGICAL MAPS

On geological maps, the outcrop pattern of folds appears as parallel belts of strata (Figure 2.13) in which the same succession of rocks is repeated. You should have already noticed that there are two belts of Carboniferous rocks on the Cheddar Sheet produced by the anticlinal structure. However, to understand the anatomy of folds in more detail you will first need to know some of the terms used to describe them.

4.4.1   Recognizing folds from their outcrop patterns

You should remember from Section 2.3.3 that, after folding, strata dip away from each other on opposite sides of an anticline and towards each other in a syncline. Therefore, when folded rocks have been eroded to expose the layers of rocks, the following is true:

> Older strata are found towards the core of an anticline, and younger strata towards the core of a syncline.

The basic geometry and terms are illustrated in Figure 4.9 which shows a series of beds deformed into an upright anticline and syncline pair: the following terms apply to both anticlines and synclines. Each fold has a limb on either side, and a **fold axis** (sometimes called a hinge line) which lies along the line of greatest curvature of the folded beds. The **axial surface** is an imaginary surface which passes through the hinge lines of successive layers and is a useful conceptual tool for determining whether a fold is *upright* or *inclined*. For instance, in upright folds, such as Figure 4.9, the axial surface is vertical, whereas in inclined folds the axial surface has a dip. Folds are also often described as being symmetrical or asymmetrical. In **symmetrical folds**, the steepness of each limb is identical (Figure 4.9); in **asymmetrical folds**, one fold limb is steeper than the other (Figure 4.10).

4.4.2   Outcrop patterns resulting from dipping fold axes

A horizontal fold axis and vertical axial plane presents us with perhaps the simplest folded outcrop pattern on a map: it results in two parallel belts of outcrop each repeating the strips representing the stratigraphic succession of the folded beds (see again Figure 2.13). However, fold axes are seldom horizontal. Where they are inclined, they are said to exhibit a **fold plunge**, the dip of which is measured in degrees below horizontal and its direction of plunge as a bearing from north. The outcrop patterns produced by these plunging fold axes are a little more complex to interpret.

Figure 4.11a–b shows an upright anticline with a fold axis gently plunging eastward, and the resulting outcrop which produces a pattern that *converges* in the direction of plunge. Figure 4.11c–d shows a westward-plunging syncline.

Figure 4.9   Cut-away block diagram illustrating the basic geometry of three beds of folded rocks comprising an anticline and syncline structure. The **hinge line** or fold axis (h), shown as dashes, joins the points of maximum curvature (m) along each of the beds. The axial surface is an imaginary plane which can be drawn through the hinge lines of the different beds. In this example the axial surfaces of both folds are vertical, so they are examples of upright folds, and because the limbs are of equal steepness, both the anticline and syncline can be described as *symmetrical*.

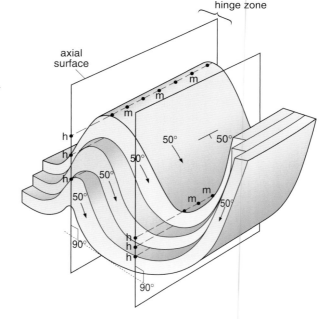

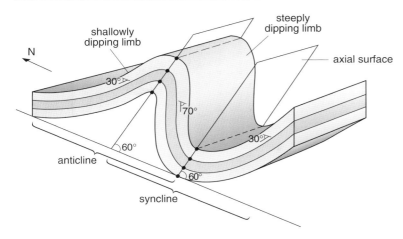

Figure 4.10 Block diagram illustrating axial surface geometry in an inclined asymmetrical anticline and syncline. Here, the axial surface dips steeply northward so the fold is *inclined*, and because the anticline has a shallowly dipping northern limb and steeply dipping southern limb it can be described as *asymmetrical*.

Here, the outcrops *diverge* in the direction of plunge. Wherever there is a plunging fold, the outcrops of the two fold limbs will converge and close, eventually meeting on the fold axis at the **fold nose** (also known as the fold closure). Adjacent anticlines and synclines often plunge in the same direction as each other so, as a result, areas of folded rocks are often characterized by zigzag outcrop patterns such as those illustrated in Figure 4.12 (overleaf).

> **Question 4.2** Given that the outcrop patterns of the eastward-plunging anticline (Figure 4.11b) and the westward-plunging syncline (Figure 4.11d) are superficially similar, how would you go about differentiating the two structures on a geological map?

A careful look at the Carboniferous strata forming the anticline on the Cheddar Sheet suggests that their outcrops converge towards the eastern part of the map. Along the eastern boundary of the map, the base of the Lower Limestone Shale on the northern strip is about 0.8 km away from its counterpart on the southern strip, whilst at eastings grid line 46 the separation has widened to about 1.5 km. Clearly, these strata are converging slightly towards the east. If you compare this outcrop pattern with that of Figure 4.11a–b, you will see that it must be created by an eastward-plunging anticline.

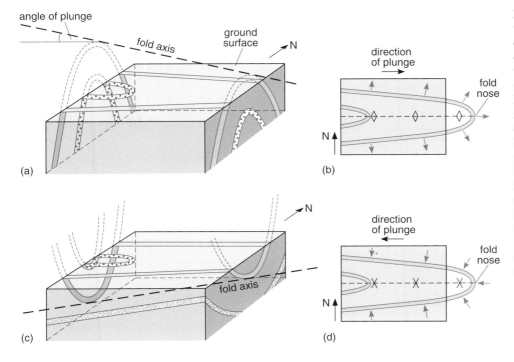

Figure 4.11 (a) A plunging anticline with the direction of plunge to the east, and (b) its corresponding outcrop pattern seen on a map. (c) A westward-plunging syncline, and (d) its corresponding outcrop pattern. Note that along the axis of a plunging fold, the bedding dip and the plunge of the fold axis are the same. Although BGS maps show neither fold axis nor plunge, when drawing sketch maps it is conventional for geologists to illustrate the direction of plunge using a large annotated arrow as shown in this Figure. Another useful symbol is to use paired arrowheads along the fold axis to show whether the beds dip toward the axis (syncline) or away from the axis (anticline).

Figure 4.12 (a) Block diagram and (b) corresponding geological map showing the outcrop pattern in an area of folded strata that plunge gently to the west. These folds should be described as upright symmetrical plunging folds (i.e. they have vertical axial plane, limbs of similar steepness and an inclined hinge axis). Harder beds in the folds form a series of zigzag ridges with steeper scarp slopes facing *inward* toward the axis of the anticline, and *outward* from the axis of the syncline. Note also the direction of the V-patterns as the rivers cut across the anticlinal and synclinal axes: V-patterns point inward toward the axis of the syncline, and outward from the axis of the anticlines.

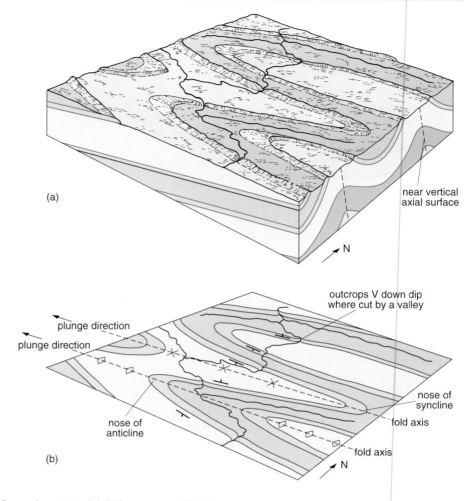

Question 4.3 (a) What type of fold is present in grid square ST 4952? (b) In which direction does it plunge?

Activity 4.3

You should now try Activity 4.3, which is a field trip to St. Andrews, Scotland, where you will be asked to make dip and strike measurements across an area of folded rock and then to investigate the outcrop patterns of folded structures using 3D computer models.

4.4.3 OTHER FOLD OUTCROP PATTERNS — DOMES AND BASINS

So far, we have considered strata which have been folded into more or less regular shapes. However, anticlines and synclines can plunge at *both* ends, in which case they are known as **domes** and **basins**, respectively: these structures can produce circular or elliptical outcrop patterns (Figure 4.13).

In Block 3, Sections 9 and 10, you will learn more detail about different types of fold, and about how folds develop.

4.4.4 OUTCROP PATTERNS RESULTING FROM OVERTURNED FOLD LIMBS

Until now, we have assumed that beds on either side of a fold axis dip in opposite directions, away from the axis in anticlines, and towards the axis in

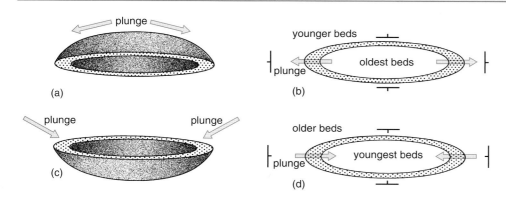

Figure 4.13   (a) An anticline can have the form of an elongate dome when the fold axis plunges in opposite directions at both ends. This produces a dome structure which, when eroded, results in an elliptical outcrop pattern (b). A syncline with its axis plunging at either end results in a basin structure (c) which also produces an elliptical outcrop pattern (d) but with the beds dipping inward toward its centre.

synclines (Section 4.4.1). However, in the case of overturned folds, this relationship becomes more complicated. An **overturned fold** is one in which one of the fold limbs has been turned upside-down (Figure 4.14a).

❑   Look at the direction of the dip arrows (i.e. those with values of 70°, 52° and 18° in square ST 4459). In which direction do these beds dip in relationship to the major anticlinal axis developed in the Devonian and Carboniferous rocks?

■   These dip arrows point southwards *towards* the anticlinal axis: this is the *opposite* direction to all the other dip arrows on the northern limb of the fold, and is the reverse to that which we have so far been led to expect on an anticlinal structure.

The reason why the dip directions are reversed in square ST 4459 is because the beds in this very small area of the fold limb have been overturned. If you look again at the dip symbols, you will see that they consist of a hooked dip arrow and a small strike line; this is the symbol used for overturned strata.

For the purposes of map interpretation, it is important to realize that when a fold limb becomes overturned, the beds on both fold limbs will dip in the same direction (Figure 4.14b), and that the beds of the overturned limb are upside-down. In cases of overturned limbs, the dip is in the opposite direction to that which you would expect from the younging direction of the strata (Section 4.3.3). Fortunately, the occurrence of overturned strata is usually identified easily on a geological map because the dip of these strata is indicated by the special symbol.

Figure 4.14   (a) Cross-section of strata deformed into an overturned fold. The axial surface is inclined and dips northward and the steeply dipping beds of the southern limb have been overturned. (Note that the eroded part of the structure is shown using dotted lines above the current erosion level.)
(b) Outcrop pattern resulting from erosion of the overturned fold. The beds of both limbs dip in the same direction (northwards), but the dips of the overturned southern limb are shown using the overturned dip symbol. Note that in the case of overturned limbs, the beds young in the opposite direction to dip. (Axial surface of fold is shown as dashed line A–A´.)

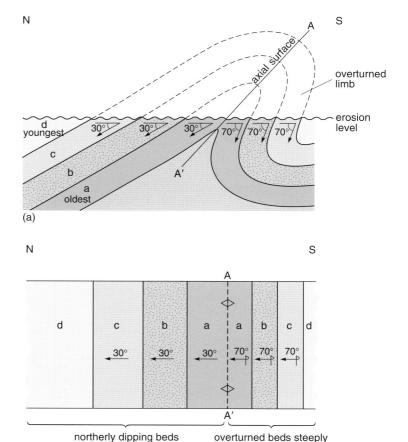

4.4.5 Determining the timing of folding of the rocks

The principles described for faulting (Section 3.4.1) also apply to folding.

> Folding is *later* than the deposition of the youngest sediment that has been folded and *earlier* than the oldest rock which is unaffected by the folding.

Since on the Cheddar map the Triassic rocks directly overlie the folded Carboniferous strata but are unaffected by the folding, we can therefore say that the folding must have occurred after the Carboniferous, but before the Triassic.

4.5 Faults

The map reveals relatively few faults in the Cheddar area. A series of normal faults cut the Jurassic rocks of the south-western corner, and a few other faults cut the small folds in pre-Triassic strata in the south-eastern area. Before we look in more detail at these faults in the south-east, we will first need to consider other different types of faults and the outcrop patterns they produce. From Section 3.4, you should remember that faults do not continue indefinitely, and most important of all, that where younger rocks lie against older rocks along a fault, the younger rocks are *always* on the downthrow side. So far, we have considered two types of fault (normal and reverse) in which mainly vertical movement in the direction of dip of the fault plane is dominant; we also briefly mentioned that thrust faults are a kind of reverse fault with a shallowly dipping fault plane.

4.5.1 Reverse and thrust faults on geological maps

Folding of rocks is the result of powerful forces which cause deformation of the Earth's crust, and reverse and thrust faults are also often associated with regions of intense folding. Nevertheless, the basic rule you learned for normal faults, that the younger rocks are always on the downthrow side, also applies to reverse and thrust faults (Figure 4.15).

Two thrust faults occur in the south-eastern corner of the Cheddar map where there is a series of synclines and anticlines. One is labelled 'south-western overthrust' (ST 477530 to ST 500512); the other is simply labelled 'thrust' and runs from ST 470534 to ST 492524. The downthrow along the south-western

Figure 4.15 (a) Block diagram of a low-angle reverse (thrust) fault in shallow dipping strata. The thrust plane dips at 30° southwards: the strata have a slightly shallower dip. Thrust faults result from deformation causing shortening of the rock strata in the horizontal plane. In this case, the thrust has carried older strata (e.g. layers 3 and 4) upwards and northwards over younger strata (e.g. layers 4 and 5). (b) Outcrop pattern resulting from the thrust after erosion has reduced topography to a level plane. Note that the thrust is shown by a toothed ornament which is the convention on many tectonic or structural maps. Note also that on the northern side of the fault line, younger rocks (layer 5) lie adjacent to older rocks (layer 3) on the southern side, therefore the downthrow must be to the north. Another important effect of the thrust is that on the plan or map view the stratigraphic succession (layers 4 and 5) is repeated on the southern side of the thrust. This also occurs in the vertical sense, since if we were to put down a borehole at locality X we would encounter layers 3 and 4 twice because they are repeated below the fault plane.

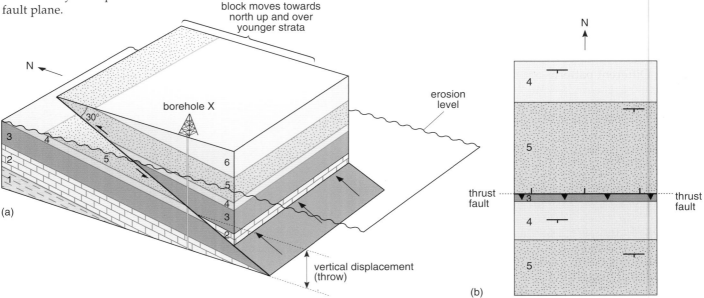

overthrust is on the north-eastern side therefore the fault plane *must* dip to the south-west. Unfortunately, since the thrust does not cross any valleys, we cannot verify this with V-patterns. As a result of the thrust, the Black Rock Limestone strata have been pushed from the south-west over the younger Hotwells Limestone. In this example, the angle of dip of the fault plane is probably not much greater than that of the strata (i.e. about 30°).

❑ Look at the succession of beds on the traverse between ST 499520 and ST 493511 which comprise the south-western limb of a north-westerly-plunging anticline. Make a list of the strata encountered along this traverse beginning with the rocks forming the core of the anticline. Do the strata on this limb always young toward the south-west, and are any of the strata repeated?

■ The axis or core of the anticline is composed of Burrington Oolite and the strata young away from the anticlinal axis (Section 4.4.1). However, further southward, beyond the south-western overthrust (i.e. ST 495515), this relationship breaks down because here older beds (i.e. Black Rock Limestone) are encountered. Moreover, further southward, the Burrington Oolite is repeated on the southernmost flank of the anticline.

Question 4.4 Suppose that an exploratory deep borehole is to be sunk at ST 484517. Knowing that the south-western overthrust extends beneath the borehole site, make a list of the succession of different beds you might expect to encounter during boring operations before you entered the Portishead Beds. (*Hint:* draw a sketch cross-section between ST 489521 and ST 480514 ; it is not necessary to include the dolomitic conglomerates (DCg).)

4.5.2 STRIKE–SLIP FAULTS

Another type of fault is the **strike–slip fault** (which elsewhere you may find called transcurrent, wrench or tear fault) in which *sideways* (i.e. horizontal) rather than vertical movements are dominant. Strike–slip faults are classified as either **dextral faults** (Figure 4.16a) or **sinistral faults** (Figure 4.16b) according to the direction of movement. Imagine you are standing on one of the blocks in Figure 4.16, facing the fault plane: if the other block moves to the right, it is a dextral fault; if the other block moves to the left, it is a sinistral fault. (*Note:* the sense of movement is the same, whichever side you view it from.)

Strike–slip faults can have the largest displacements of all faults. The displacement across these faults can be hundreds of kilometres and the slip is primarily in a horizontal direction parallel to the strike of the fault.

4.5.3 OUTCROP PATTERNS PRODUCED BY FAULTING

It is useful to consider faults in terms of the effect they have on the pattern of outcrop. Different patterns are produced depending upon whether the fault cuts across or is parallel to bedding strike. A **dip-parallel fault** is where the outcrop of the fault plane is roughly parallel to the dip direction of the beds, therefore it must cut across the bedding strike. Faults oriented in this fashion result in an 'offset' pattern of beds at the surface. Figure 4.17a–c illustrates the pattern resulting from the movement on a normal fault which cuts a series of steeply dipping strata. A similar 'offset' pattern is produced where a reverse fault cuts across the bedding strike (Figure 4.18a–c). Since the two offset patterns are the same when presented on a geological map, it is not possible to determine whether the offset has been caused by a normal or a reverse fault from simply observing the outcrop patterns. In these situations, additional information indicating the downthrow side (i.e. tick marks, V-patterns, or other field observations that help indicate the direction of dip of the fault plane) would normally be required to distinguish normal and reverse faults. It is also very important to note that in the cases of both the normal and reverse faults, the

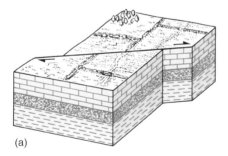

(a)

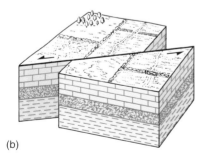

(b)

Figure 4.16 Block diagrams of strike–slip faults. (a) Dextral: movement of far block is to the right. (b) Sinistral: movement of far block is to the left.

Figure 4.17 Pattern of rock outcrop resulting from a normal fault (F–F′) perpendicular to the strike of dipping beds: (a) prior to fault movement; (b) vertical movement offsetting beds (wavy line on upthrown block shows surface after erosion); (c) outcrop pattern (i.e. map view) resulting after erosion of both blocks to a common topographic level.

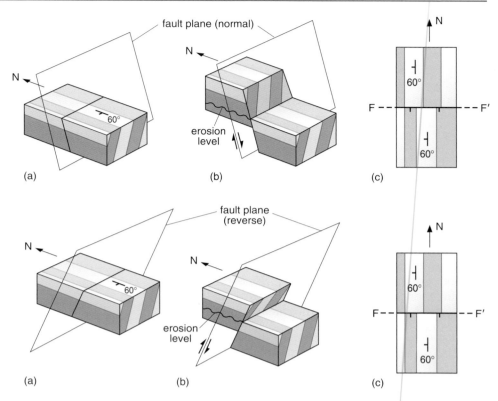

Figure 4.18 Pattern of rock outcrop resulting from a reverse fault (F–F′) perpendicular to the strike of dipping beds: (a) prior to fault movement; (b) vertical movement offsetting beds (wavy line on upthrown block shows surface after erosion); (c) outcrop pattern (i.e. map view) resulting after erosion of both blocks to a common topographic level.

'offset pattern' of the dipping strata has been produced by a predominantly vertical motion (described as a dip–slip fault). *No* sideways (i.e. strike–slip) movement has occurred on the fault despite the fact that the bed outcrop pattern *appears* to be offset.

Let us now consider the types of outcrop pattern that would result from a fault cutting the dipping strata in a strike-parallel orientation (i.e. a **strike-parallel fault**). Here, the important factors which fundamentally affect the outcrop pattern are the angle of dip of the fault plane, and whether the fault plane dips in the same direction as the stratigraphic dip, or opposite to it. Instead of producing an offset outcrop pattern, the outcrop width of the dipping beds cut by the fault may apparently be altered, whilst in some cases beds may be repeated at the surface, or else removed altogether. A good example of the type of repetition and narrowing of outcrop that can result from a strike-parallel reverse fault is given in Figure 4.15b.

❑ How might the outcrop pattern appear after erosion if the amount of vertical movement (throw) shown in Figure 4.15a had been half as much again as the amount illustrated? (A simple sketch of the outcrop pattern may help.)

■ Since a greater amount of layer 4 would have been removed by erosion, we would expect to see a wider sliver of layer 3 lying immediately south of the fault. Traversing from north to south across the two blocks, we would therefore expect to encounter beds 4, 5, the fault, a wider outcrop of 3, and then 4 and 5 repeated further southward on the southern, upthrown block.

It is most important to realize that commonly, in many real geological situations, faults are neither perpendicular nor parallel to strike, and most cut the strata at an angle that lies somewhere between the dip and strike directions.

Activity 4.4

Now attempt Activity 4.4, which enables you further to explore how different fault orientations might affect surface outcrop patterns in areas of dipping or folded strata.

Having completed Activity 4.4, you should realize that it is possible to generate similar outcrop patterns by either vertical or horizontal displacements and, therefore, that it can become very difficult to determine the type of fault by simply looking at the outcrop patterns. In cases of faults where we have no other information (e.g. tick marks which would then show that vertical motion is dominant, and that we are dealing with either normal or reverse faults), we can only be certain that dominantly horizontal (i.e. strike–slip) movement has taken place *if* the fault is crossed by, and displaces, a vertical feature such as a dyke, a vertical bed, or an upright fold. Only where vertical features are demonstrably displaced by faults does it become possible to measure on the map just how much lateral displacement has occurred.

In Activity 4.4, you will have directly encountered the different relationships between outcrop pattern and type of faulting. The important differences in outcrop which should enable you to distinguish strike–slip and dip–slip faults are as follows:

- Dip–slip faults (i.e. normal or reverse faults) typically have steeply dipping (but not usually vertical) fault planes. When strata dipping in opposite directions (e.g. fold limbs) are cut by dip–slip faults, the predominantly vertical movements produce an outcrop pattern where geological boundaries are displaced in opposite directions. However, boundaries of vertical geological features such as dykes will show no displacement by dip–slip movement.

- Strike–slip faults are typically less common on BGS maps than dip–slip faults. The fault plane is usually vertical (or nearly so) and so its trace cuts straight across topography. Displacement of geological boundaries cut by the fault will be in the same direction on the same side of the fault regardless of dip direction and amount.

Generally, unless you can find definite evidence of strike–slip motion (i.e. vertical features or fold limbs displaced in the same direction), it is safe to assume most faults you will see are either normal or, less commonly, reverse.

4.6 UNCONFORMITIES

In Section 4.1.3, we identified the presence of an unconformity in the generalized vertical section, and determined that it represents a major gap in the stratigraphic record of the Cheddar area. The vertical section shows no record of Carboniferous strata younger than the Quartzitic Sandstone Group, or of any Permian age strata. Therefore, either these were never laid down, or they were eroded prior to the deposition of the Triassic succession (i.e. Keuper Marl and Dolomitic Conglomerate).

If you now look at the dip arrows of the Triassic and Jurassic succession cropping out in the southern half of the map, you will see that they generally have a much lower dip (about 3–5°) than the outcrops of Carboniferous and Old Red Sandstones (ORS) which form the anticline to the north. Further inspection also reveals that Triassic rocks cross over the boundaries between different Carboniferous strata: this relationship can be seen best where 'tongues' of Dolomitic Conglomerate extend across different steeply dipping Carboniferous strata on the northern limb of the anticline (e.g. ST 490580, ST 476584 and ST 465588 lie at the southern tip of three such tongues). This type of outcrop relationship, together with the differences in dip angles, suggest that the Carboniferous and ORS rocks must have been folded, uplifted and eroded prior to the deposition of the Triassic and Jurassic beds above the unconformity. Moreover, the fact that there is *angular discordance* between the bedding planes in the rocks above and below the erosion surface (Figure 4.19) means that the Cheddar unconformity must be an **angular unconformity** (Section 2.4.2).

Figure 4.19   Sketch cross-section along the valley from ST 476581 to ST 475594. (Note the major difference in the dips above and below the unconformity and also that the presence of an unconformity is illustrated by a wavy line.)

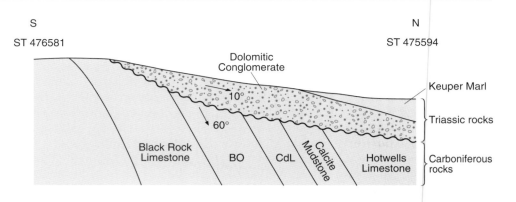

Angular unconformities are recognized in the field by their angular discordance between two vertically adjacent stratigraphic packages. The cross-section on the Cheddar Sheet shows a large angular discordance between the Carboniferous and Triassic strata at the northern end of the cross-section and a small angular discordance between them in the south where the dip of the Carboniferous is much less.

So how did the Cheddar unconformity develop? Above the unconformity, marine fossils are not found until the beginning of the Jurassic, so we must assume a non-marine origin, and most of the Triassic sediments are interpreted as having been deposited sub-aerially (i.e. on land) under semi-arid or even desert-type conditions. We know that the folding of the Carboniferous and ORS strata must have produced a range of hills much larger than the Mendips of today: these probably stood above a surrounding low-lying desert area. We also know that over much of the Cheddar area, the rocks at the base of the Triassic consist of a conglomerate, especially in the south-eastern part of the map.

The Triassic 'tongues' represent the infilling of valleys carved in the hills of Carboniferous rocks by torrential streams produced through sporadic heavy rainfall during the Permian and earlier parts of the Triassic Period. This rainfall must have also caused sudden flooding called flash floods, and the valleys and lower slopes at the edge of the uplands became buried and infilled by coarse-grained sediment containing angular fragments of Upper Palaeozoic sediments (i.e. Carboniferous and ORS material) which today form the tongues of 'Dolomitic Conglomerate'. Away from the hills, finer-grained sediments were deposited on the desert plain to form the Keuper Marl. Much later, during the late Triassic and early Jurassic Periods, a relative rise in sea-level caused flooding of this desert plain, and the marine limestones of the White and Blue Lias were deposited (Figure 4.20).

Since the Mendips have again been uplifted and today form an upland area, much as they did in Triassic times (Figure 4.21), they are again being eroded. This erosion is preferentially attacking and removing the conglomerates which infilled the Triassic valleys and is thus slowly exposing the ancient **buried**

Figure 4.20   Diagrammatic cross-section of a non-marine erosional unconformity similar to that of the Mendips. The topography beneath the unconformity is very irregular and the hills and valleys have been progressively infilled by screes and conglomerates adjacent to the higher land, and sands and gravels in the more distant lowlands. Finer-grained sediments become more common as topography becomes progressively buried over time.

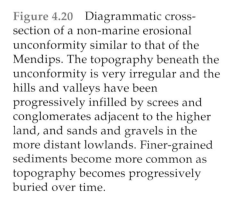

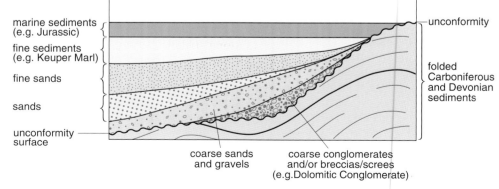

(a)

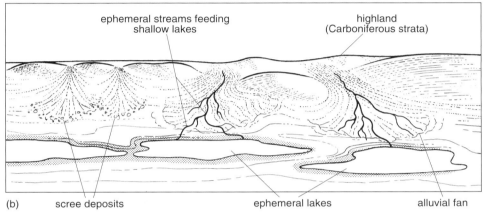

(b)

Figure 4.21 (a) The Mendips today, viewed from the south. The plains in the middle ground are underlain by Triassic sediments. (IPR/12-14 British Geological Survey © NERC. All rights reserved.) (b) An impression of how the Mendips may have appeared during Triassic times if viewed from the same locality. The finer sediments that underlie the plain are just beginning to accumulate; alluvial fans and scree deposits are spilling down onto a low-lying alluvial plain.

topography. From inspection of the contours, you can see that the tongues of Dolomitic Conglomerate coincide with modern valley features (e.g. ST 4155, ST 4855 and ST 4659). In such cases where sediments above irregular unconformities are eroded so exposing the ancient valleys and hills buried beneath them, it is customary to speak of an **exhumed topography**. The unconformity here is thus a highly irregular surface, rather than the nearly planar marine unconformity you met in Section 2.4.2.

4.6.1 IDENTIFYING UNCONFORMITIES ON GEOLOGICAL MAPS

In your examination of the Carboniferous strata on the Cheddar map, your attention has already been directed toward the fact that the Triassic sediments immediately above the unconformity form numerous 'tongues' that cut across the boundaries of older rocks and cover over substantial areas of them. These were produced when Triassic sediments buried an existing landscape of substantial relief comprising folded Carboniferous and ORS strata (Figures 4.19 and 4.20). This can be readily appreciated if you look again at the three tongues of Triassic rocks at ST 490580, ST 476584 and ST 465588. It is clear that *under* the Triassic (i.e. below the unconformity), the Burrington Oolite and Clifton Down Limestone must connect with each other along strike. Therefore, wherever you see this type of pattern where younger beds cut across and obscure boundaries between older strata, you should at once suspect the presence of an unconformity above a buried topography with tongues of younger sediments representing valley infills.

In Section 2.4.2, we also mentioned that inliers and outliers are commonly associated with unconformities. In the north-western and south-eastern corners of the map, there are several inliers of Carboniferous rock which are hills and ridges surrounded by the blanketing cover of younger Triassic sediments. For

Box 4.1 Distinguishing between faults and unconformities on a geological map

At first glance, you may find it difficult to distinguish the outcrop pattern produced by an unconformity from that produced by faults. So how do we identify unconformities on geological maps? The main distinguishing features are summarized by the following checklist:

• The presence and stratigraphic position of major unconformities on any IGS/BGS map sheet is typically shown on the generalized vertical section.

• Angular unconformities have an obviously irregular, often curving, outcrop pattern which cuts across and partially obscures the outcrop patterns of older strata.

• Stratigraphic boundaries in the succession below the unconformity are commonly truncated by the unconformity.

• In the case of angular unconformities, the dip of the overlying unconformable strata is significantly different (usually less) than that of the strata beneath.

• On BGS maps, unconformities are depicted as normal geological boundaries, whereas faults are commonly shown as heavy black broken lines which

tend to displace adjacent strata usually along *straight* or *gently curved* lines.

• Some gently dipping thrust faults will V across valleys and thus may appear like an unconformity. However, the important distinction here is that a thrust puts older rocks on top of younger ones, whereas an unconformity has younger ones above older ones (often with a significant age gap between).

• The only other circumstance where difficulty in identifying an unconformity might arise is where an unconformity surface has been steeply inclined by later tilting or folding. In such cases, it may resemble a fault when plotted on a geological map. Again, the difference in line ornament should help you to distinguish between the two.

• Importantly, the crucial information about the presence of unconformities is also often available from the stratigraphic information printed with most geological maps. Any periods of geological time not represented in the succession of strata should become apparent by examining the generalized vertical section. In many cases, such gaps are actually marked as 'unconformities'.

example, at Nyland Hill (ST 458504) and Banwell Camp (ST 410590) the hill base is surrounded by conglomerate, in a similar fashion to that depicted on Figure 4.20. Box 4.1 summarizes how to distinguish between faults and unconformities on maps.

4.7 CROSS-SECTIONS IN TILTED AND FOLDED STRATA AND ACROSS UNCONFORMITIES

Sometimes it is useful to draw a **sketch cross-section** before attempting an accurate one. The purpose of this is to give you a 'feel' for the geological structure of an area, and it is quite acceptable to draw it rather crudely since it is intended to help guide *your* understanding of the geological relationships (you may have already constructed one when answering Questions 3.12 and 4.4). Consider for instance how the section printed on the Cheddar Sheet would have looked if it had been drawn so that its northern end were about 200 m further to the west, i.e. it would have intersected the Triassic at the head of the valley at ST 476584, and run down the valley to Langford Green (ST 475594). Figure 4.19 is a sketch cross-section along this line to show you what is happening under the unconformity; remember that on sketch cross-sections it is normal to illustrate the position of unconformities using a wavy line (e.g. see Figures 4.19 and 4.20).

In Activity 4.5, you will complete an *accurate* cross-section. However, before you begin, go through the checklist of preliminaries in Box 4.2.

Box 4.2 Preparing to draw an accurate cross-section

• How does the line of the proposed cross-section relate to the cross-section provided with the map itself? If the two lines cross or are very close, use the printed information to aid you on the appropriate part of your section.

• If there is no cross-section provided, then determine how the line of the proposed cross-section relates to the *strike* of the strata.

• On some maps you will not find any useful dip arrows, so you have to deduce directions of strike and dip from outcrop patterns. If the proposed line of section is almost at right angles to the strike, *that is a very important finding*, because it means that you can transfer dip angles from the map onto the cross-section (provided your section has similar vertical and horizontal scales). When the line of cross-section is *not* perpendicular to the strike then, on your cross-section, you have to show an *apparent* dip which must always be less than that of the true dip (Section 4.3.1). Look all along the line of the proposed cross-section. Does the dip of the strata vary in different places?

• Are there faults? If so, determine and indicate their downthrow sides. In which directions do the fault planes dip? Make a note of all these features.

• Is there an unconformity present in the stratigraphic succession? Check carefully the presence of unconformable relationships by examining outcrop patterns and information given in the generalized vertical section. Note them on your cross-section.

• When drawing sections you should draw in all the beds which are above the unconformity *first*. Next, it is often useful to put in one bed or bedding plane on both limbs of any fold present to show the overall structure before filling in all the details of the section.

• Use annotation where necessary to distinguish the different units. Mark on axial surfaces of folds and label other important geological features such as faults and unconformities. Dashed lines can be used to complete fold structures above ground and helps make the overall structure clearer.

You should also note the following before beginning the Activity and attempting Question 4.5:

• This section runs almost parallel to the cross-section below the map.

• You know that the strata dip north in the northern part and south in the southern part of the map, and that dips are steeper in the north than in the south. Somewhere in between, they must be horizontal, which is near where the fold axis is to be found. The line of section is almost at right angles to the strike, so you can transfer dip angles directly from the map to the section.

• Moreover, you know about the unconformity, which is overlain by nearly horizontal Triassic sediments that obscure some of the dipping Carboniferous strata beneath. There are no faults along the line of section, but there are numerous mineral veins which may well occupy fault planes. No dip or downthrow information can be gleaned from these; so, just for the purposes of constructing the cross-section, assume that the veins are almost vertical and that any displacements along them are not measurable at this scale.

Question 4.5 (a) Using the Cheddar Sheet, draw a sketch map of the first anticline lying north-east of the south-western overthrust (ST 4952) marking the fold axis. In which direction does the fold axis plunge?

(b) Draw a cross-section between ST 490524 and ST 500521 (i.e. along the anticlinal axis). (*Note:* to avoid complication, keep your line of cross-section south of the thrust fault affecting the Clifton Down Limestone near the fold nose. Assume the fold plunges at about 10°.)

Activity 4.5

Now do Activity 4.5 in which you will construct an accurate geological cross-section.

4.8 GEOLOGICAL HISTORY OF THE CHEDDAR AREA

In Section 3.6, you were shown how to work out the sequence of geological events that had occurred in the Moreton-in-Marsh area. You can apply similar principles to the Cheddar area, as shown in Box 4.3; this checklist is longer than the one in Activity 3.2 because we are now dealing with a more complex area. You can use a similar checklist for other complex areas.

Box 4.3 Compiling a geological history

• Find the oldest sediments on the map. Their deposition is the first event of the geological history.

• Is there an unconformity present? If so, it is normally shown in the stratigraphic column on the map.

• Do the beds below the unconformity show any of the following: (a) folding; (b) faulting; (c) igneous intrusions; (d) metamorphism?

• Is there evidence that these events did *not* affect rocks above the unconformity (i.e. that they occurred before the younger strata were deposited)? For example, a fault may cut older strata but disappear at the unconformity. If more than one of (a)–(d) has occurred, work out the order in which they happened by determining which features (younger) *cut across* the others (older).

• Uplift above sea-level and erosion before deposition of the next series of beds form the next event of the geological history (i.e. the formation of the surface of unconformity).

• Is there another, later, unconformity present? If so, repeat steps 3, 4 and 5 until the latest series of sediments is reached.

• Have the most recent sediments been: (a) folded; (b) faulted; (c) intruded; (d) metamorphosed?

• What evidence is there for tilting and erosion since the area was uplifted to form land?

• What evidence is there of any glacial activity (e.g. presence of Boulder Clay)?

• What evidence is there of more recent deposits such as river terrace deposits?

The main principle to be applied here is that *younger* features tend to *cut across* older features; these are also known as *cross-cutting relationships*. For example, faults and intrusive igneous rocks can affect only those strata which are older than the time during which faulting or intrusion occurred; beds above a plane of unconformity cut across (i.e. cover) the older eroded strata below the unconformity, and recent deposits such as Boulder Clay cut across ('blanket out') all earlier geological features.

 Question 4.6   Outline the sequence of geological events which have affected the area in grid square ST 4654.

4.8 OBJECTIVES FOR SECTION 4

Now you have completed this Section, you should be able to:

4.1   Identify strike and dip directions on maps.

4.2   Understand the difference between true dip and apparent dip and how these affect the dip of strata on cross-sections that are not perpendicular to strike.

4.3   Identify dip directions on maps from the stratigraphic succession, and use outcrop patterns including V-ing strata.

4.4.   Explain the difference between outcrop width and true unit thickness in areas of dipping strata, and perform simple calculations relating these parameters.

4.5   Identify simple folds on maps from their outcrop patterns and the stratigraphic succession.

4.6   Distinguish between upright and inclined, symmetrical and asymmetrical folds and identify the direction of any fold plunge.

4.7   Identify different types of faults cutting tilted and folded strata, and recognize the downthrow side.

4.8   Identify unconformities on geological maps.

4.9   Construct accurate sketch geological cross-sections across maps showing simple arrangements of tilted and folded strata and unconformities; and make measurements of bed thickness and fault throws on such sections.

4.10 Interpret the geological history of an area containing folded strata and unconformities.

Now try the following questions to test your understanding of Section 4. You will first need the Cheddar Sheet for Question 4.7, and then the Ten Mile Maps (N & S) for Questions 4.8–4.11.

Question 4.7 On the southern side of the main anticline:

(a) In terms of difference in ages, what is the relationship between the Quaternary (Pleistocene and Recent) Head deposits, and the immediately underlying Mesozoic strata?

(b) How do you explain the fact that in many cases the outcrop of these head deposits seems to follow the outcrop tongues of Triassic strata (e.g. in square ST 4554)?

Question 4.8 On the Ten Mile Map (S), what evidence is there for the direction of dip of the Jurassic rocks where they cross the Humber estuary (SE (44) 92)?

Question 4.9 (a) On the Ten Mile Map (S) how would you describe the base of the Permian and Triassic sandstones (89) between Shrewsbury (SJ (33) 5014) and just east of Wrexham (SJ (33) 3450)?

(b) Is the Cretaceous (102–6) north of the Humber (TA (54) 0025) separated by an unconformity from the rocks below? Give reasons for your answer.

Question 4.10 On the Ten Mile Map (N), look at the outcrops of Carboniferous rocks comprising Millstone Grit (81) and Coal Measures (82–83) around Dalkeith (NT (36) 3268) lying just east of Edinburgh.

(a) On what type of fold does Dalkeith lie?

(b) Does this fold have a plunge and, if so, in which direction?

(c) Where do (i) Dalkeith and (ii) Lasswade (NT (36) 3067) lie with respect to the fold axis?

Now look at the same succession of Carboniferous rocks exposed on the northern side of the Firth of Forth around NT (36) 39.

(d) Is there any evidence that the Carboniferous rocks of this area form part of a fold, and if so what is the approximate direction of plunge?

(e) Using the information given on the Ten Mile Map (N), what type of structure would you find if you were to predict the outcrop pattern of Millstone Grit and Coal Measures extending beneath the Firth of Forth between the towns of Buckhaven on the north (NT (36) 3798) and Musselburgh (NT (36) 3573) on the south? (*Hint:* A sketch map of the Carboniferous sediments of this area may help.)

Question 4.11 On the Ten Mile Map (N), look at the area 5–10 km south-west of Wigton (NY (35) 2548) where the Carboniferous and Permian rocks (82–87) display a saw-tooth outcrop pattern against the Permian and Triassic (89); this is caused by a series of NW- to SE-trending faults.

(a) Are these dip-parallel or strike-parallel faults?

(b) Do these faults all have the same direction of throw?

(c) Which is the downthrow side of these faults?

(d) What is this type of faulting called?

Question 4.12 How would you record the following strike and dip measurements in a field notebook?

(a) A bed with a strike exactly east–west, and a dip of 30° northward.

(b) A bed dipping exactly south-west with a dip of 50°.

(c) A strike of 010° and a dip of 70°.

ANSWERS TO QUESTIONS

Question 1.1

You should have concluded that the Outer Hebrides and north-west Scotland consist of the oldest rocks in Britain with early Precambrian (Lewisian) rocks, and that the youngest rocks are the Cenozoic clays and sands of the London and Hampshire areas.

Question 1.2

You should have written down the stratigraphic column for the whole of the Cenozoic, Mesozoic and Palaeozoic, with the exception of the Cambrian Period, which is not encountered along this line. You may also have crossed a small area of Silurian rocks between the Carboniferous and the Devonian in the vicinity of the Scottish border, and the last few kilometres to Edinburgh are on Carboniferous rocks after crossing the oldest rocks of the traverse, namely the Ordovician.

Question 1.3

Metamorphic rocks are almost entirely confined to the Highlands of Scotland and north-west Ireland. You may also have discovered small areas in Anglesey and the Lleyn Peninsula in North Wales. There is also a very small patch in South Devon (Start Point).

Question 1.4

The largest extent of igneous intrusive rocks is in the Highlands of Scotland, so they appear to be associated mainly with metamorphic rocks. There are also igneous intrusions in northern and south-western England and various parts of Ireland, with some small occurrences in West Wales. The volcanic igneous rocks show a rather different distribution. They are more extensively associated with sedimentary rocks, but again are concentrated in Scotland, northern and north-western England, North Wales and Northern Ireland.

Question 1.5

See Figure A1.1 (opposite).

Question 1.6

It should be apparent that virtually all the mountainous and high ground of Britain is underlain by Palaeozoic or older rocks or by metamorphic or igneous rocks. This is because these rocks are generally much harder and more resistant to erosion than Mesozoic and Cenozoic sedimentary rocks.

Question 1.7

(a) Dorchester.

(b) Chalk.

(c) Phanerozoic, Mesozoic and Cretaceous.

(d) Between 142 and 65 million years ago according to the time-scale on Figure 1.2.

Question 1.8

(a) Permian/Triassic.

(b) Lower Cambrian.

(c) Lower Cambrian (64), Lower Carboniferous basal conglomerate (79), Carboniferous limestone (80), Permian Magnesian limestone (86), and Permian and Triassic sandstones (89) and Triassic mudstones (90).

Figure A1.1   Answer to Question 1.5.

Question 1.9

(a) Both areas of rock belong to the Devonian Period and are between 354 and 417 million years old.

(b) London lies on the Tertiary and marine early Quaternary (1.6–65 million years) and Dublin on the Carboniferous (290–354 million years). The maximum age difference possible is about 352 million years if London were on the youngest Tertiary beds and Dublin on the oldest Carboniferous.

Question 2.1

(a) The high ground dominates the north-western section of the map around the mountains Bla Bheinn and Garbh Bheinn. There is also high ground to the north-east of the map extract around the mountain called Beinn na Caillich. The Strathaird Peninsula forms a lower area of land, although the western side of it has a couple of higher areas around Ben Cleat and Ben Meabost. Most of the lower ground is on the east side of the Strathaird Peninsula and around the village of Torrin (NG 580 210).

(b) The road north-east from Elgol parallels the contours of the map for the majority of its path. There are no steep gradients, although there is a marked dip in the road around Kilmarie (NG 546175) where the road crosses an inlet.

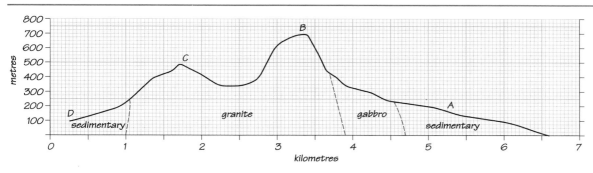

(c)   The flattest areas are around the south-east corner of the Strathaird Peninsula in the area to the east of Elgol (NG 530 140).

Question 2.2

See Figure A2.1.

Question 2.3

(a) 28% (1 in 3.6); (b) 13% (1 in 7.5); (c) 27% (1 in 3.8).

Question 2.4

The soft Kimmeridge Clay Formation can form cliffs because the top of the cliff is formed by the more erosion-resistant Portland Beds. At the foot of the cliff, vigorous wave action is undermining the mudstone; the mudstone higher up tumbles down to beach level and is then washed away by the sea. The Portland Beds are also undermined by this process, but they fall in massive blocks which cannot be removed as quickly by the sea.

Question 2.5

The topographic profile is shown in Figure A2.2.

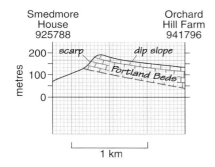

Figure A2.2 (Answer to Question 2.5) Topographic profile of Smedmore Hill. The probable base of the Portland Beds has been sketched in. Note that the vertical scale has been exaggerated to give a × 3 vertical exaggeration, thus all dips are represented as being steeper than they actually are.

Question 2.6

The contours 1.5 km north of Gad cliff show another scarp slope trending roughly east to west, parallel to the Portland Scarp. Therefore, it is probable that the northern scarp slope is formed by another relatively erosion-resistant tilted stratum above the Portland Beds.

Question 2.7

At both ends of the Isle of Wight the strata must be dipping to the north, because younger strata occur in that direction: 104 in the south and 109 in the north.

Question 2.8

Where the width of the Chalk (106) outcrop widens, it must be dipping more gently. Thus, at the Needles its outcrop is very narrow since the dip is almost vertical. To the east, the outcrop widens as the dip decreases near Newport. Still further east, the dip steepens again and the width decreases once more.

Question 2.9

(a)   The contour interval on this map is 25 feet; the heavier contour lines are at 100 foot intervals.

(b)   The lowest land is in the south-west of the map around Ingleton.

(c)   The highest point on the map is Ingleborough Hill at 2372 feet, which is in the eastern part of the map.

(d)   There are two main rock features depicted: (i) linear cliffs along the valley sides (often called scars, e.g. Raven Scar), shown by a symbol usually used for quarry edges (contours are close together, indicating very steep slopes); and (ii) areas of rock outcrop above the cliffs e.g. Green Ridge (7375) and Ewes Top (7076). Here, the contours are spaced more widely indicating that the slopes are less steep.

Question 2.10

Around Twistleton Scar End, the contours are parallel to the cliffs. This tells us that the beds which form the cliffs must be roughly horizontal here. If the beds were dipping, the cliffs would not continue at the same height around this spur.

Question 2.11

See Figure A2.3 (overleaf). The average gradient (a) between P and Q is 1 in 3.6, and (b) between Q and R is 1 in 25.

Question 2.12

Most potholes are between the 900 foot and 1500 foot contours, see especially squares 7172 and 7272. Gaping Hill Hole is at about 1335 feet; Meregill Hole is at about 1265 feet.

Question 2.13

(a)   The general strike direction is east–west.

(b)   Older beds occur going south from London, from 108 to 102. They must therefore dip to the north.

(c)   Southwards from East Grinstead younger beds occur. They must be dipping to the south.

(d)   The structure is an anticline. See Figure A2.4.

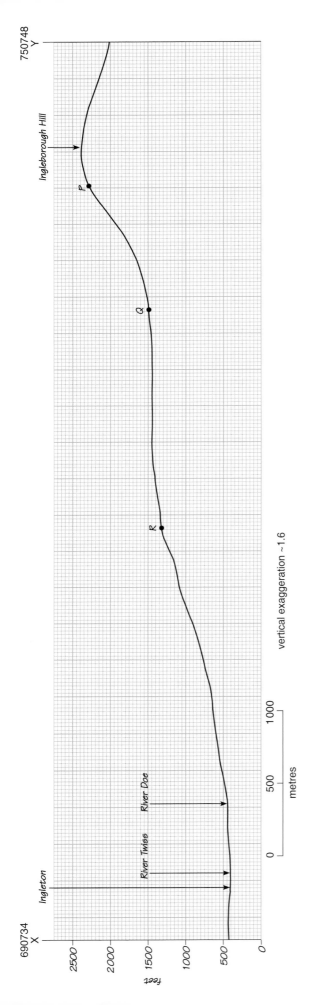

Figure A2.3   Answer to Question 2.11.

Question 2.14

See Figure A2.5 (opposite).

Question 2.15

(a)   The sequence is Carboniferous Limestone (80), Millstone Grit Series (81), Coal Measures (82), Pennant Measures (83), i.e. younger rocks to the north. Therefore, the strata are dipping to the north.

(b)   Just south of the River Usk, the sequence is reversed and older strata occur to the north. Therefore, the strata are dipping to the south.

(c)   The structure is a syncline, with younger rocks (83) in the centre.

(d)   The strike direction is east–west, as shown by the east–west stripes of the outcrops between Ammanford and Brynmawr.

(e)   The strike direction is again east–west, as shown by the outcrop pattern between Margam and Caerphilly.

(f)   The strike direction is roughly north–south, given by the outcrop pattern.

(g)   Dip must be either to the east or to the west (i.e. at right angles to the strike direction). The map tells us that younger strata occur from east to west (from bed 75 in the east to bed 83 in the west), so the dip is to the west.

Question 2.16

(a)   Since the strata are dipping, harder rocks will form scarp features and dip slopes.

(b)   The scarps will be outward-facing, i.e. facing to the north on the northern part of the syncline and to the south on the southern part of the fold. See Figure A2.6 (opposite).

Question 2.17

The outcrop pattern on the northern limb of the syncline is, in general, wider than that on the south. Two possible reasons are:

(a)   Change in thickness of the beds. It is possible that all the beds are thinner in the south than in the north.

(b)   Change in dip. It is likely that the difference in outcrop widths is due to a difference in the angle of dip between the northern and southern limbs. The northern limb will be at a shallower angle than the southern limb.

Question 2.18

Strike is north–south, dip is to the west.

Question 2.19

(a)   Permian and Triassic rocks (89,90).

(b)   Cambrian (64–66), Carboniferous (82–84) and Permian (89).

(c)   The beds in the west are the younger ones, so these must have been moved down relative to the older strata in the east.

Question 3.1

The downthrow is to the north, where the (younger) Middle and Upper Lias (MLi and ULi) have been downthrown against the (older) Lower Lias (LLi) which lies on the southern side of the fault. (*Note:* there is no special reason why a tick mark is not shown on this particular fault.)

Question 3.2

The top of the Fuller's Earth to the south has been downthrown to rest opposite the top of the Cotteswold Sands to the north. In

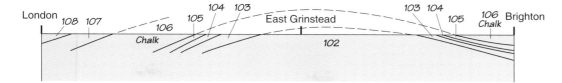

Figure A2.4 Answer to Question 2.13(d).

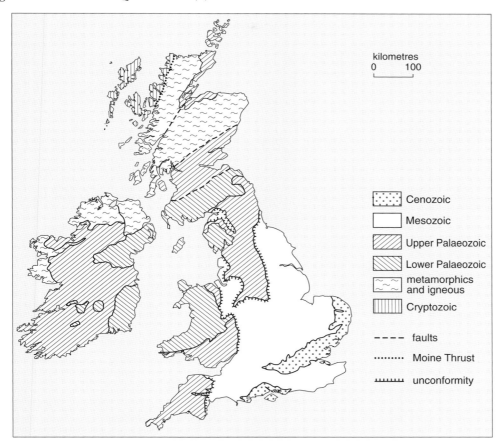

Figure A2.5 Answer to Question 2.14.

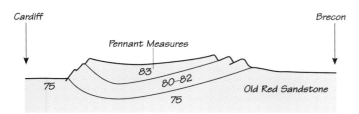

Figure A2.6 Answer to Question 2.16(b).

the pre-faulted situation, the vertical distance between the top of the Csd (i.e. the base of the InO) and the top of the FE would have been equivalent to the full thickness of the InO plus that of the FE (i.e. 50 m + 10 m = 60 m). After faulting the top of the Csd and top of the FE are juxtaposed either side of the fault, so the fault's vertical displacement (i.e. throw) must be 60 m. Figure A3.1 illustrates this displacement in terms of a stratigraphic column.

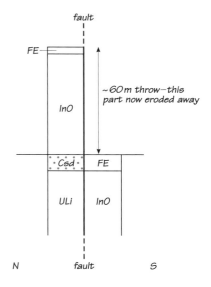

Figure A3.1 For use with Question 3.2.

Question 3.3

A similar approach can be adopted to that of Question 3.2. The maximum thickness of the InO is about 80 m and that of the FE is 11m from the generalized vertical section; this gives a total of 91 m for the maximum possible throw. The minimum thicknesses are 15 m and 0 m, respectively, which gives a minimum possible throw > 15 m since clearly a thickness of 0 m for the FE is unreasonable because the unit is definitely present at this locality.

Question 3.4

See Figure A3.2. This cross-section shows a faulted outlier which is the result of a graben developing between two east–west faults. *IMPORTANT:* You may find that your geological cross-sections do not exactly match ours in every detail. This is no cause for concern: the main point is for you to understand the procedure for constructing cross-sections in an area of near-horizontal strata and correctly indicating the relative thicknesses of beds.

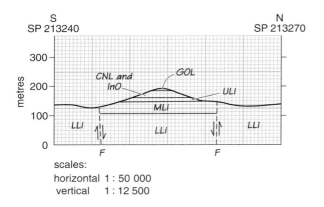

Figure A3.2 For use with Question 3.4.

Question 3.5

(a) See the cross-section in Figure A3.3. Across the section, there is a very slight westerly dip on the strata.

(b) There are two possible answers to this question, both of which are correct. It is clear from tick marks and the geological relationships that the north and south boundaries of this tract of rocks are normal faults both of which throw down the central region. This can, therefore, be accurately described as a *graben*. However, close inspection around the margins of the feature reveals that the stratigraphic succession within this graben is

everywhere surrounded by older rocks (ignoring the river alluvium). To the north and south, this relationship is the result of the faulting, whilst to the east and west it is the result of erosion. Therefore it would also be accurate to call this region a *fault-bounded outlier*.

Question 3.6

(a) Middle Lias.

(b) The road *climbs* uphill (the contours increase in value) over *younger* strata (from Middle Lias to Inferior Oolite).

(c) There are two main lines of evidence. First, there are several symbols for horizontal strata (+) located in the vicinity. Secondly, the boundaries between the strata closely follow the contour lines (e.g. the InO/Csd junction between SP 153370 and SP 140375 runs parallel to the 213 m contour line).

Question 3.7

(a) In both cases the MLi and ULi form outcrops which are entirely surrounded by older LLi rocks.

(b) The Cheltenham Sands are part of the unconsolidated Quaternary succession (i.e. drift deposits). Moreover, they may only ever have been deposited as isolated patches so their distribution may not be the result of either erosion or faulting as we would expect for a true outlier.

(c) The tract of geology comprising FE and GOL (Stonesfield Slate) is bounded to the north and south by two normal faults. Since these faults have downfaulted the FE and GOL relative to the surrounding InO, this structure can be described as a *graben*. The pattern of outcrop resulting from this graben is that of a fault-bounded outlier.

Question 3.8

(a) The downthrow side is to the south. Younger MLi and ULi on the south of the fault lie against older LLi and MLi respectively on the northern side.

(b) (i) At SP 013351, the boundary between LLi and MLi on the north side lies adjacent to the boundary between the MLi and ULi units on the south side. In other words, the base of the Middle Lias unit on the north lies against the top of the Middle Lias on the south. According to the stratigraphic column, the MLi is between 18 m and 75 m thick so this gives us a maximum and minimum value for the throw. (*Note:* we can assume that the thickness of the Marlstone Rock Bed, MRB, is zero since it is absent at this locality.) However, since strata are flat-lying, we can use contours to help determine the thickness of strata. In-

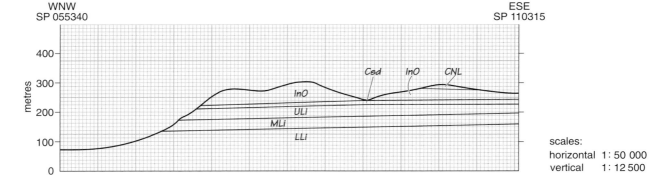

Figure A3.3 For use with Question 3.5.

spection of the contours around this hill (i.e. how many contours occur within the outcrop of the MLi) indicates that the MLi is of the order of 40–50 m thick here, so in fact this latter value would give us a more accurate determination of the throw.

(ii) Since the fault tip dies out here, the throw must be approaching zero. Compare this situation with Figure 3.10.

(c) The fault throws the younger CNL on the eastern side downwards against the older InO located on the western side of the fault. The downthrow is therefore to the east.

Question 3.9

(a) The downthrow is to the north-west where the younger Lower Lias (91) and Middle Lias (92) occur juxtaposed against older Triassic (90) rocks.

(b) There are no geological boundaries on the south-eastern side of the fault to match up with the Triassic/Lias (90/91) and Lower Middle Lias (91/92) boundaries which run into the fault from the north-west. Even if suitable boundaries did occur, there is no thickness information on this map sheet with which to determine fault displacement.

(c) These are outliers because they are surrounded by older strata.

Question 3.10

At the north-eastern end of the Bala fault, the Permian–Triassic sandstones (89) cover the tip of the fault, indicating that any fault movement must have occurred prior to deposition of these sediments. In other words, the fault must be older than Permian. The fault itself cuts strata of Cambrian through to Carboniferous; the youngest strata affected are of Upper Carboniferous (Westphalian; 82 & 84) age. Therefore the last movement must have occurred after deposition of the Upper Carboniferous strata, but before deposition of the Permian–Triassic sandstones.

Question 3.11

(a) The outcrop is an inlier because the Carboniferous Limestone Series (80) is surrounded by the younger rocks of the Millstone Grit Series (81).

(b) The outcrop of Coal Measures (82–83) is an outlier because it is surrounded by older rocks of the Millstone Grit Series (81).

Question 3.12

See Figure A3.4.

Question 4.1

To calculate true thickness:

t = outcrop width × sine of dip (average = 28.5°).

(a) Width of outcrop on the map is 27 mm. Horizontal scale of the map is 1 : 25 000, so 1 mm represents 25 m. Therefore, width of outcrop on the ground is 27 × 25 = 675 m.

$$t = 675\,\text{m} \times \sin 28.5°$$
$$t = 675\,\text{m} \times 0.4772 = 322\,\text{m}$$

Therefore, approximate true thickness ≈ 320 m.

(b) Width of map outcrop is 12 mm, or 300 m on the ground, and averaged dip is 75°.

$$t = 300\,\text{m} \times \sin 75°$$
$$t = 300\,\text{m} \times 0.9659 = 289\,\text{m}$$

Therefore, approximate true thickness ≈ 290 m.

Comment: First, note that because we only measured the outcrop width to two significant figures, we would be claiming unrealistically high precision if we were to quote a greater number of significant figures in our answer. Therefore, we round off our result to two significant figures. Secondly, the maximum thickness for the Black Rock Limestone (BRL) shown on the generalized vertical section is about 270 m, but becomes thinner where it is capped by dolomite. The two thicknesses calculated above are consistent with this observation, because the BRL appears thinner to the east and west of Dolebury Warren where it is overlain by dolomite in places, and is thicker at ST 462562 where the dolomite appears totally absent. However, the two calculated thicknesses are both *greater* than the value indicated on the generalized vertical section. The most probable explanation is that the averaged dip values we have taken from the map are an overestimate of the real dip values. If we had instead used shallower dip values, then we would be closer to the thickness of the BRL indicated on the generalized vertical section. The fact that the dip angle we used may not be accurate or fully representative is another reason for caution.

Question 4.2

If dip arrows are given, then the strata forming an anticline will dip away from the fold axis, while those of a syncline will dip towards the fold axis. In the absence of dip data, the anticline will contain the oldest strata in the centre (core) along the fold axis; a syncline will have the youngest strata exposed in its core. If valleys had been eroded into the folded strata, then another method would be to examine the outcrop V-patterns. If you remembered that outcrop Vs of dipping strata on geological maps tend to point in the direction of dip, you should then be able to work out whether the fold outcrop pattern was that of an anticline or syncline (see Figures 4.4 and 4.12).

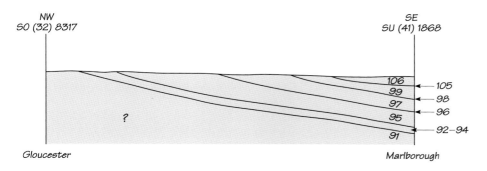

Figure A3.4 For use with Question 3.12.

Question 4.3

(a) The dip arrows, and outcrops of the older beds (e.g. Burrington Oolite) in the core of the fold, show that this fold is an anticline.

(b) The outcrop pattern reveals that the fold nose lies to the north-west: therefore it plunges to the north-west (cf. Figure 4.11b).

Question 4.4

The borehole site is in the centre of a gently folded syncline comprising the Burrington Oolite (BO) and Black Rock Limestone (Figure A4.1). The thrust has pushed this shallow syncline up the side of the anticline that lies to the north-east (Question 4.3). The drill would first go through the Burrington Oolite, Black Rock Limestone and Lower Limestone Shale of the syncline. It would then pass through the thrust plane where it would encounter the succession comprising the southern flank of the anticline beginning with the Hotwells Limestone then downward through the stratigraphic succession that we can see exposed in the core of the anticline.

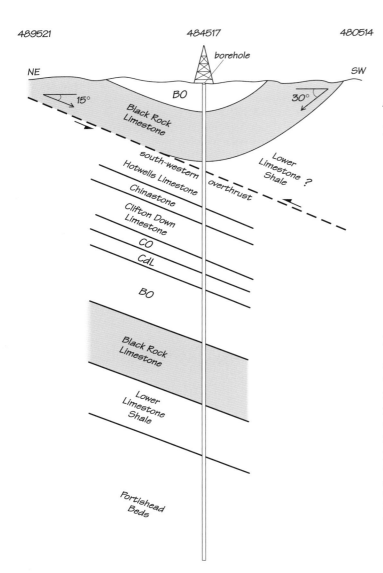

Figure A4.1 For use with Question 4.4.

Question 4.5

(a) See Figure A4.2. From the outcrop pattern, it is evident that this anticline plunges to the north-west since this is the direction of closure of the fold nose.

(b) See Figure A4.3.

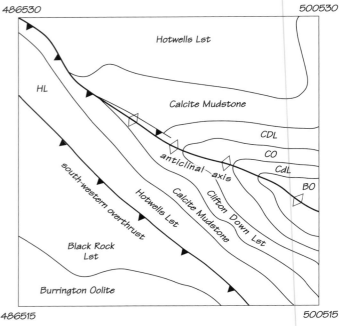

Figure A4.2 For use with Question 4.5(a).

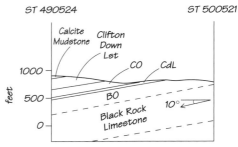

Figure A4.3 For use with Question 4.5(b).

Comment: When constructing the sketch cross-section, although we know the direction of plunge of the fold axis, there is in fact no direct evidence given as to its angle of plunge. However, a rough estimate of fold axis plunge (i.e. whether it is likely to be steep or shallow) can be made from the map information. Directly along the axis of a plunging fold, the bedding dip and plunge of the fold axis are the same (see again Figure 4.11a, b), so if we work out the dip of the beds lying exactly along the fold axis (i.e. *along the line of your section*) we should find the angle of plunge. The bedding dip can be calculated from the width of the Cheddar Limestone cropping out along the line of section by using the equation in Section 4.3.5 which relates true (stratigraphic) thickness (*t*), width of outcrop (*w*) and dip (θ):

$$\sin \theta = \frac{t}{w}$$

We can measure the width of outcrop on the cross-section or the Cheddar map (i.e. 6 mm × 25 000 = 150 m), and from the generalized vertical section we can make a rough estimate of 20 m for the average vertical thickness of the Cheddar Limestone (it varies between 0–30 m thickness).

Therefore,

$$\sin \theta = \frac{20 \text{ m}}{150 \text{ m}} \approx 7-8°.$$

Question 4.6

The sequence of geological events (geological history) of this small area can be summarized as follows:

1 Conformable deposition under marine conditions of the Clifton Down Limestone succession (here incorporating beds of the Cheddar Limestone, Cheddar Oolite and Clifton Down Limestone) during Early Carboniferous times possibly succeeded by younger Carboniferous beds now lost to erosion. This succession rests on an earlier succession of beds including Portishead Beds (Devonian) and Lower Limestone Shale, Black Rock Limestone and Burrington Oolite (Carboniferous) that do not outcrop at the surface in this grid square but can be inferred at depth.

2 Folding (and associated thrusting) of the strata, followed by uplift and erosion, took place prior to the deposition of the next sedimentary succession which occurred during Triassic times. This means that the folding probably occurred in Late Carboniferous or Permian times.

3 Deposition of the Dolomitic Conglomerate unconformably upon the eroded Carboniferous strata. In this square, the Triassic rocks are confined to a tongue of outcrop which sits in the bottom of a valley feature.

4 Continuing erosion of the area leads to preferential removal of the Dolomitic Conglomerate and exhumation of a post-Carboniferous erosion valley.

5 Deposition of Head during melting of ice at the end of the last Ice Age. This material lies directly upon both Mesozoic and Palaeozoic rocks at the foot of the hillslope.

6 Continuing erosion resulting in features comprising the present landscape.

Question 4.7

(a) Inspection of the geological column indicates a considerable length of time (i.e. more than 200 Ma) elapsed between deposition of the Triassic rocks and these Pleistocene deposits.

(b) Many of the outcrops of Head on the southern slopes of the Mendips on the Cheddar Sheet follow a similar pattern to that of the Triassic. Since Head represents material which has moved or slid downhill, we can assume that the Permian valleys were already partially exhumed before the Head began to accumulate. The Head simply began to refill these valleys during Pleistocene times.

Question 4.8

The beds dip eastward because younger Jurassic rocks lie toward the east. Also, there is an eastward V in their outcrop pattern as they cross the Humber estuary.

Question 4.9

(a) It is an unconformity because the base of bed 89 cuts across the boundaries between the older beds. For instance, it lies upon the Upper Carboniferous (84) at Shrewsbury, then passes across Silurian (73–4) and Ordovician (70) going west, then north onto Lower Carboniferous (80), and finally back on to Upper Carboniferous (84) near Wrexham.

(b) Yes, because going north from the Humber, it initially lies on Upper Jurassic (98–9), then passes on to progressively lower beds down to the Lower Lias at Market Weighton (SE (44) 8842), and then back on to Upper Jurassic inland of Filey (TA (54) 1280).

Question 4.10

(a) Dalkeith lies on a syncline (younger rocks occur in the centre of the structure).

(b) The outcrop of Coal Measures rocks (82–3) trends NNE–SSW, and the fold nose of the syncline lies at the southern end; this, together with the younging direction, indicates that the fold must plunge toward the north-east.

(c) (i) Dalkeith lies almost directly upon the fold axis.

(ii) Lasswade lies at the boundary between the Millstone Grit and Coal Measures on the north-western limb of the syncline (i.e. north-west of the fold axis).

(d) A pattern of outcrop similar to that of the Dalkeith syncline can be recognized because Coal Measures strata exposed along the coastline (82–83) are surrounded by Millstone Grit strata. This outcrop pattern is consistent with a broad syncline plunging toward the south-east.

(e) We can make a prediction of the geology beneath the Firth of Forth by joining up the corresponding geological boundaries marking the contact between the Carboniferous Limestone Series (80) and the Millstone Grit (81), and the Millstone Grit and the Coal Measures (82–83) cropping out on the northern and southern shores. When this is done, the outcrop pattern links the two plunging synclines to reveal a single synclinal structure with a curved axis plunging at *both* ends. Another name for this type of structure is a *basin* (see Figures 4.13c, d and A4.4.)

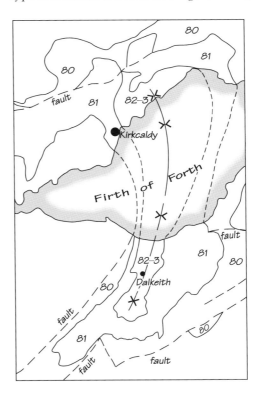

Figure A4.4 Sketch map (for Question 4.10(e)) showing Carboniferous sediments that have been deformed into a basin-type structure beneath the Firth of Forth.

Question 4.11

(a) Where they cut the Triassic they are more like dip-parallel faults than strike-parallel faults because the general strike of the strata is WSW to ENE and the faults are perpendicular to this direction.

(b) Yes.

(c) Downthrow is to the north-east since younger beds crop out on this side.

(d) Step faulting, where a series of sub-parallel normal faults all have a similar direction of downthrow (Section 3.4, Figure 3.8d).

Question 4.12

(a) 090/30N (alternatively 270/30N).

(b) 135/50S (alternatively 315/50S, or even 135/50 SW).

(c) It is impossible from the information given to tell whether the dip is toward the east (e.g. 010/70E), or the west (e.g. 010/70W): this example illustrates the importance of always noting the approximate direction of dip in order to avoid confusion.

ACKNOWLEDGEMENTS

Grateful acknowledgement is made to the following for permission to reproduce material in this Block:

Cover image copyright © Derek Hall; Frank Lane Picture Agency/Corbis; *Figure 1.4a* Ordnance Survey; *Figures 1.5 and 1.6* courtesy of the Geological Society; *Figure 2.6* Aerofilms Ltd.; *Figure 4.21a* reproduced by permission of the British Geological Survey © NERC. All rights reserved.

INDEX

Note: page numbers in **bold** are for terms that appear in the *Glossary* while page numbers in *italics* are for terms contained within Figures.